THE
OMEGA POINT

BANTAM NEW AGE BOOKS

This important imprint includes books in a variety of fields and disciplines and deals with the search for meaning, growth and change. They are books that circumscribe our times and our future.

Ask your bookseller for the books you have missed.

THE
OMEGA POINT

The search for the missing mass and the ultimate fate of the Universe

JOHN GRIBBIN

BANTAM BOOKS
TORONTO • NEW YORK • LONDON • SYDNEY • AUCKLAND

THE OMEGA POINT

A Bantam Book / August 1988

New Age and the accompanying figure design as well as the statement "the search for meaning, growth and change" are trademarks of Bantam Books, a division of Bantam Doubleday Dell Publishing Group, Inc.

Library of Congress Cataloging-in-Publication Data

Gribbin, John R.
The Omega Point.

Bibliography.
Includes index.
1. Cosmology. 2. Astrophysics. I. Title.
QB981.G76   1987   523.1'9   87-47908
ISBN 0-553-34515-X

Published simultaneously in the United States and Canada

Bantam Books are published by Bantam Books, a division of Bantam Doubleday Dell Publishing Group, Inc. Its trademark, consisting of the words "Bantam Books" and the portrayal of a rooster, is Registered in U.S. Patent and Trademark Office and in other countries. Marca Registrada. Bantam Books, 666 Fifth Avenue, New York, New York 10103.

PRINTED IN THE UNITED STATES OF AMERICA

O   0 9 8 7 6 5 4 3 2 1

The law that entropy always increases – the second law of thermodynamics – holds, I think, the supreme position among the laws of Nature. If someone points out to you that your pet theory of the universe is in disagreement with Maxwell's equations – then so much the worse for Maxwell's equations. If it is found to be contradicted by observation – well, these experimentalists do bungle things sometimes. But if your theory is found to be against the second law of thermodynamics I can give you no hope; there is nothing for it but to collapse in deepest humiliation.

Arthur Eddington,
The Nature of the Physical World

Some say the world will end in fire,
Some say in ice,
From what I've tasted of desire
I hold with those who favor fire . . .

Robert Frost,
Fire and Ice

ACKNOWLEDGEMENTS

In writing about new developments at the cutting edge of research, I have been indebted to many astronomers and physicists for their willingness to talk about their new ideas. The following people, in particular (listed in no special order) helped through discussions, providing copies of their published papers and/or advance copies of papers not yet published, and, in some cases, correcting my misconceptions. Thanks to: John Huchra, Tom Kibble, Roger Tayler, Carlos Frenck, Vera Rubin, Frank Tipler, John Barrow, Michael Rowan-Robinson, Stephen Hawking, Jim Peebles, David Wilkinson, John Faulkner, John Ellis, Marcus Chown, Tjeerd van Albada, Adrian Melott, John Bahcall, Willy Fowler, Ron Gilliland, William Press and Lawrence Hall. Any remaining misconceptions are, of course, all my own work.

CONTENTS

INTRODUCTION

Where are we going? The ultimate fate of humankind, and of the Universe we inhabit, holds a fascination which has fuelled many a myth, fired religious fervour since time immemorial, and provided philosophers with food for thought down the centuries. Some have regarded the world as essentially unchanging, a stage on which the actors may come and go but which is itself eternal: others have built upon the idea of repetition and cycles in the Universe, with changes occurring, but only, ultimately, to bring everything back to its original starting condition, ready for another cycle, perhaps the same as before, or perhaps with minor (or major) variations. For most of human history, the third possibility, of a Universe which was born in a unique creation event, and will end, in a finite time, in a unique moment of destruction, has been very much a minority view. But that possibility is now at the forefront of scientific debate.

Cosmologists, the scientists who specialise in the study of the Universe at large, can best explain the nature of the Universe as we see it today in terms of a definite beginning, a moment of creation that they call the Big Bang, which occurred some 15 billion years ago. The Universe of matter and energy, space and time, appeared in a superdense state, according to their calculations, and has been expanding and thinning out ever since. And the same equations which so successfully describe the birth of the Universe offer only two choices for its ultimate fate. Depending upon just how much mass there is in the entire observable Universe, it will either continue to expand forever, getting thinner and darker as stars age and die, or the expansion will one day be halted and reversed, producing a contraction that will lead ultimately to a big crunch, the mirror image of the big bang of creation.

"Big crunch" is, however, an ugly term which hardly seems appropriate for so important an event as the end of the Universe. Indeed, "big bang" is a rather inappropriate term for the moment of creation, and was invented by Fred Hoyle, who has never accepted the idea, as a term of derision. It seems to have stuck, to the chagrin of some cosmologists, and I shan't fly in the face of convention by rejecting the name. But there is no convention as yet for a label of the moment of destruction at the end of time, and I am free to borrow, from the writings of the French-born Jesuit philosopher Pierre Teilhard de Chardin, the term which gives me the title of this book – The Omega Point.

Teilhard, of course, used the term in a rather different context. He had no real interest in the Universe at large, but was concerned with the spiritual evolution of consciousness, through humankind, to some ultimate state, his Omega Point, named after the last letter of the Greek alphabet. The philosophy, expressed in his book *The Phenomenon of Man*,* is not to my taste – though not for the same reasons as those of the Roman Catholic Church, which, in 1962, issued a warning against uncritical acceptance of Teilhard's ideas. To them, I suspect, Teilhard was being too materialistic in his attempts to incorporate evolutionary ideas into Christian faith; to me, the ideas are nowhere near materialistic enough, ignoring as they do the vast bulk of the Universe! But he did describe his Omega Point in terms that would almost be familiar to modern cosmologists:

---

* Full details of sources mentioned can be found in the Bibliography. The quote on the following page comes from page 259 of *The Phenomenon of Man*.

> Spacetime ... must somewhere ahead become in-
> voluted to a point which we might call *Omega*, which
> fuses and consumes [everything] integrally in itself.

This, indeed, is a description of the mirror image of the moment of creation described by big bang theory. That theory of the origin of the Universe is, perhaps, the greatest achievement of scientific thought. It is a story that seems to be largely complete (although you can never be sure that any scientific idea may not be overturned by new and even better ideas, this one certainly marks, at the very least, the end of a major phase of scientific thinking about our origins), and which I reported on in detail in my book *In Search of the Big Bang*. So, increasingly, cosmologists are turning their attention to the other half of the puzzle, the question of the ultimate fate of the Universe. Since that depends on the amount of matter in the Universe, the vital element in the story is, and will be, the search for all the different kinds of matter in the Universe.

The bright stars are, of course, the obvious constituents of our Universe but astronomers have known for decades that there must be much more material around in the form of dark matter, and the question now is how much of this dark matter there might be, what form it might take, and where it may be hiding in the depths of space. By a happy coincidence, there *is* a convention among astronomers for the label which they attach to this critical parameter. The density of matter in the Universe is denoted in cosmology by Ω, the capital Greek omega! It is defined in such a way that if the cosmological omega is less than one, the Universe will expand forever, while if it is bigger than one then we are heading inevitably for the big crunch, the omega point.

Astronomers have not yet detected enough matter to do the trick, but that doesn't mean that it is not there. With a fondness for alliteration, some astronomers call this as yet unobserved material the "missing mass", although really it is the *light* that is missing, and some dark matter is certainly out there in some form or another. It's a catchy name, which I shall borrow, with apologies to those of my astronomical friends who prefer the more scrupulous accuracy of the term "missing light". But is there enough missing mass to ensure that the Universe ends in the omega point?

The search for the big bang was a largely twentieth-century scientific adventure, but one that is now almost over; the search for the missing mass, and thereby for an understanding of the ultimate fate of the Universe, is a story that is only just beginning. Where my earlier book gave the best scientific answers to questions about the origin of the Universe, now I can only offer you the best *questions* that cosmologists and others are asking each other, and indicate what sort of answers they hope to find. There may be a few loose ends in the story, and ideas which fall by the wayside in the next few years. But this is not scientific history, it is science in the making – and that, in some ways, is even more exciting.

CHAPTER ONE

THE ARROW OF TIME

The most important feature of our world is that night follows day. The dark night sky shows us that the Universe at large is a cold and empty place, in which are scattered a few bright, hot objects, the stars. The brightness of day shows that we live in an unusual part of the Universe, close to one of those stars, our Sun, a source of energy which streams across space to the Earth and beyond. The simple observation that night follows day reveals some of the most fundamental aspects of the nature of the Universe, and of the relationship between life and the Universe.

If the Universe had existed for an eternity, and had always contained the same number of stars and galaxies as it does today, distributed in more or less the same way throughout space, it could not possibly present the appearance that we observe. Stars pouring out their energy, in the form of light, for eternity, would have filled up the space between themselves with light, and the whole sky would blaze with the brightness of the Sun. The fact that the sky is dark at night is evidence that the Universe we live in is changing, and has not always been as it is today. Stars and galaxies have *not* existed for an eternity, but have come into existence relatively recently; there has not

been time for them to fill the gaps in between with light.
Astrophysicists, who study the way in which the stars
produce their energy, by nuclear reactions deep in their
hearts, can also calculate how much light a typical star
can pour out into space during its lifetime. The supply of
nuclear fuel is limited, and the amount of energy a star
can produce, essentially by the conversion of hydrogen into
helium, is also limited. Even when all the stars in all the
galaxies in the known Universe have run through their
life-cycles and become no more than cooling embers, space,
and the night sky, will still be dark. There is not enough
energy available to make enough light to brighten the
night sky. The oddity, the strangeness of the observation
that night follows day, is not that the sky is dark, but that
it should contain *any* bright stars at all. How did the Universe come to contain these short-lived (by cosmological
standards) beacons in the dark?

That puzzle is brought home with full force by the light
of the Sun in the daytime. This represents an imbalance in
the Universe, a situation in which there is a local deviation
from equilibrium. It is a fundamental feature of the world
that things tend towards equilibrium. If an ice-cube is
placed in a cup of hot coffee, the liquid cools and the ice
melts as it warms up. Eventually, we are left with a cup of
lukewarm liquid, all at the same temperature, in equilibrium. The Sun, born in a state which stores a large
amount of energy in a small volume of material, is busily
doing much the same thing, giving up its store of energy
to warm the Universe (by a minute amount) and, eventually, cooling into a cinder in equilibrium with the cold of
space. But "eventually", for a star like the Sun, involves a
time span of several thousand million (several billion)
years; during that time, life is able to exist on our planet
(and presumably on countless other planets circling
countless other stars) by feeding off the flow of energy out
into the void.

Because night follows day, we know that there are
pockets of non-equilibrium conditions in the Universe. Life
depends on the existence of those pockets. We know that
the Universe is changing, because it cannot always have
existed in the state we observe today and still have a dark

sky. The Universe as we know it was born, and will die. And so we know, from this simple observation, that there is a direction of time, an arrow pointing the way from the cosmological past into the cosmological future.

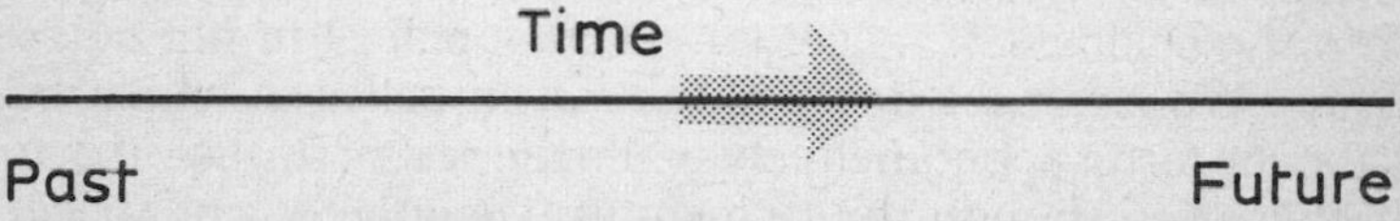

Time

Past   **Future**

*Figure 1.1/*The Arrow of Time
In the everyday world, many events proceed only in one direction, as things age or wear out. We represent this, conventionally, as an arrow of time. But this arrow should only be thought of as implying that there is an asymmetry in the world, a difference between "past" and "future". It tells us nothing about motion "through" time, which is a deeper mystery.

THE SUPREME LAW

All these features of the Universe are bound up with what Arthur Eddington, a great British astronomer of the 1920s and 1930s, called the supreme law of Nature. It is named the second law of thermodynamics, and was discovered, during the nineteenth century, not by astronomical studies of the Universe but from very practical investigations of the efficiency of the machines that were so important during the Industrial Revolution – steam engines.

It may seem odd that such an exalted rule of nature should be the "second" law of anything; but the first law of thermodynamics is simply a kind of throat-clearing statement, to the effect that heat is a form of energy, that work and heat are interchangeable, but that the total amount of energy in a closed system is always conserved – for example, if our coffee cup is a perfect insulator, once the ice-cube has been dropped into the hot coffee, although the ice warms and the coffee cools, the total energy inside the

cup stays the same. This in itself was an important realisation to the pioneers of the Industrial Revolution, but the
second law goes much further.*

There are many different ways of stating the second law,
but they have to do with the features of the Universe that I
have already mentioned. A star like the Sun pours out heat
into the coldness of space; an ice-cube placed in hot liquid
melts. We never see a cup of lukewarm coffee in which an
ice-cube forms spontaneously while the rest of the liquid gets
hotter, even though the two states (ice-cube + hot coffee)
and (lukewarm coffee) contain exactly the same amount of
energy. Heat *always* flows from a hotter object to a cooler
one, never from the cooler to the hotter. Although the
amount of energy is conserved, the distribution of energy
can only change in certain ways, irreversibly. Photons
(particles of light) do not emerge from the depths of
space to converge on the Sun in just the right way to heat

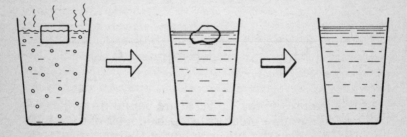

Figure 1.2/Heat always tends to even out. An ice cube
placed in a beaker of hot liquid melts, and the liquid
cools. We *never* see ice cubes form spontaneously out
of cold liquid, while the remaining liquid heats up. This
is the second law of thermodynamics, which is related
to the arrow of time.

---

* There is also a "zeroth" law of thermodynamics, an afterthought put in by a
later generation of scientists, which concerns the definition of temperature; for
my purpose here, we can manage quite happily with the everyday understanding
of temperature as a measure of hotness. The third law of thermodynamics has to
do with the impossibility of matter ever cooling to the ultimate low temperature,
the absolute zero (just under $-273°C$). Again, these subtleties are not important
here. Of the four laws of thermodynamics, the second is the one that is important to an understanding of the evolution of the Universe.

it up and drive the nuclear reactions in its core in reverse.

Stated like this, it is clear that the second law of thermodynamics also defines an arrow of time, and that this is the *same* arrow as the arrow of time defined by the observation of the dark night sky. Another definition of the second law involves the idea of information – when things change, there is a natural tendency for them to become more disordered, less structured. There is a structure in the system (ice-cube + hot coffee) that is lost in the system (lukewarm coffee). In everyday terms, things wear out. Wind and weather crumble stone and reduce abandoned houses to piles of rubble; they never conspire to create a neat brick wall out of debris. Physicists can describe this feature of nature mathematically, using a concept called entropy, which we can best think of as a *negative* measure of information, or of complexity.* *Decreasing* order in a system corresponds to *increasing* entropy. The second law says that in any closed system, entropy always *increases* (or, at best, stays the same) while complexity *decreases*.

The concept of entropy helps to provide the neatest, and best, version of the second law, but one which is only really useful to mathematical physicists. Rudolf Clausius, a German physicist who was one of the pioneers of thermodynamics, summed up the first and second laws in 1865: the energy of the world is constant; the entropy of the world is increasing. Equally succinctly, some unknown modern wit has put it in everyday language: You can't get something for nothing; you can't even break even. This is apposite because entropy, and the second law, can also be thought of as telling us something about the availability of *useful* energy in the world. Peter Atkins, in his excellent book *The Second Law*, points out that since energy is conserved, there can hardly be an energy "crisis" in the sense that we are using up energy. When we burn oil or coal we simply turn one (useful) form of energy into another (less useful, less concentrated) form. Along the way, we increase the *entropy* of the Universe, and diminish the

---

* To the mathematical physicists, it is information, or order, that is described in terms of negative entropy, or as "negentropy".

quality of the energy. What we are really faced with is not an energy crisis, but an entropy crisis.

Life, of course, seems to be an exception to this rule of increasing entropy. Living things – a tree, a jellyfish, a human being – take simple chemical elements and compounds and rearrange them into complex structures, highly ordered. But they are only able to do so by using energy that comes, ultimately, from the Sun. The Earth, let alone an individual living thing on Earth, is *not* a closed system. The Sun is constantly pouring out high-grade energy into the void; life on Earth captures some of it (even coal and oil are stored forms of solar energy, captured by living things millions of years ago), and uses it to create complexity, returning low-grade energy to the Universe. The local decrease in entropy represented by the life of a human being, a flower, or an ant is more than compensated for by the vast increase in entropy represented by the Sun's activity in producing the energy on which that living thing depends. Taking the Solar System as a whole, entropy *is* always increasing.

The whole Universe – which must, by definition, be a closed system in this sense of the term – is in the same boat. Concentrated, "useful" energy inside stars is being poured out and spread thin throughout space, where it can do no good. There is a struggle between gravity, which pulls stars together and provides the energy which heats them inside to the point where nuclear fusion begins, and thermodynamics, seeking to smooth out the distribution of energy in accordance with the second law. As we shall see, the story of the Universe is the story of that struggle between gravity and thermodynamics. When or if the whole Universe is at a uniform temperature, there can be no change, because there will be no net flow of heat from one place to another. Unless it contains enough matter to ensure collapse into the omega point, that will be the fate of our Universe. There will be no order left in the Universe, simply a uniform chaos in which processes like those which have produced life on Earth are impossible. Many scientists of the nineteenth century – and even later – worried about this "heat death" of the Universe, an end implicit in the laws of thermodynamics. None seem to have appreciated fully that the corollary of the changes we see going on in the Universe is that there must have been a birth, a "heat birth", at some finite time in the past, which created the non-equilibrium conditions we see today. And all would surely have been astonished to learn that in all but the most trivial detail the "heat death" has already occurred.

LIGHT AND THERMODYNAMICS

Energy at high temperature is low in entropy, and can easily be made to do useful work. Energy at low temperature is high in entropy and cannot easily be made to do work. This is straightforward to understand, since energy flows from a hotter object to a cooler one, and it is easy to find a cooler object than, say, the surface of the Sun, into which energy from the Sun can be made to flow and do work along the way. It is hard to find an object colder than – say – an ice cube, so that we can extract heat from the ice-cube and use it to do work. On Earth, it is much more likely that heat will flow *in* to the ice-cube. Things would be a little different in space, where it is much colder than the surface of the Earth. An ice-cube at 0 °C would still contain some useful energy which could be extracted and made to do work under those conditions. But still, there is a limit, an absolute zero of temperature, 0 K on the Kelvin scale named after another of the thermodynamic pioneers. An object at 0 K contains no heat energy at all.*

Space itself is not quite as cold as 0 K. The energy that

---

* Thermodynamic temperature is a measure of the amount of heat energy a body possesses. There is a fundamental zero point of thermodynamic temperature, equivalent to the minimum amount of heat energy a body can have. It corresponds to −273 °C. Thermodynamic temperature is measured in units called kelvins, abbreviated as K (with no "degree sign") from this zero point. Ice melts at 273 K; water boils at 373 K.

fills the space between the stars is in the form of electromagnetic radiation, or photons. The energy of these
photons can be described in terms of temperature – sun-
light contains energetic high temperature photons, while
the heat radiated by your body is in the form of lower
energy, cooler photons, and so on. One of the greatest
discoveries of experimental science was made in the 1960s,
when radio astronomers found a weak hiss of radio noise
coming uniformly from all directions in space. They called
it the cosmic background radiation; the hiss recorded by
our radio telescopes is produced by a sea of photons, with
a temperature of only 3 K, that is thought to fill the entire
Universe.

This discovery was the single key fact that persuaded
cosmologists that the big bang theory is a good description
of the Universe in which we live. Studies of distant galaxies
had already shown that the Universe today is expanding,
with clusters of galaxies moving further apart from one
another as time passes. By imagining this process wound
backwards in time, some theorists had argued that the
Universe must have been born in a superdense, superhot
state, the fireball of the big bang. But the suggestion did
not meet with general acceptance until the discovery of
the background radiation, which was quickly interpreted
as the leftover radiation from the big bang fireball itself.*

According to the now standard view of the birth of the
Universe, during the big bang itself the Universe was filled
with very hot photons, a sea of highly energetic radiation.
As the Universe has expanded, this radiation has cooled,
in exactly the same way that a gas cools when it is allowed
to expand into a large empty volume (which is the basic
process which keeps the inside of your refrigerator cool).
When a gas is compressed, it gets hot – you can feel the
process at work when you use a bicycle pump. When a
gas expands, it cools. And the same rule applies if the "gas"
is actually a sea of photons.

During the fireball stage of the big bang, the sky *was*
ablaze with light throughout the Universe, but the ex-

---

* Details of all this can be found in *In Search of the Big Bang*.

pansion has cooled the radiation all the way down to 3 K (the same expansion effect helps to weaken starlight, but not enough to explain the darkness of the sky if the Universe were infinitely old). The amount of everyday matter in the Universe is very small, and the volume of space between the stars and galaxies is very large. There are many, many more photons in the Universe than there are atoms, and almost all of the entropy of the Universe is in those cold photons of the background radiation. Because those photons are so cold, they have a very high entropy, and the addition of the relatively small number of photons escaping from stars today is not going to increase it by very much more. This is why it is true to say that the heat death of the Universe has already occurred, in going from the cosmic fireball of the big bang to the cold darkness of the night sky today; we live in a Universe that has very nearly reached maximum entropy already, and the low entropy bubble represented by the Sun is far from being typical.

The expansion of the Universe also provides us with an arrow of time – still pointing in the same direction – from the hot past to the cold future. But there is something rather odd about all this. An arrow of time, change and decay are fundamental features of the Universe at large and of everyday things that we are used to on Earth – on what physicists call a macroscopic scale. But when we look at the world of the very small, atoms and particles (what physicists call the microscopic world, although we are really talking of things far too small to see even with a microscope), there is no sign of a fundamental time asymmetry in the laws of physics. Those laws "work" just as well in either direction, forwards and backwards in time. How can this be reconciled with the obvious fact that time flows, and things wear out, in the macroscopic world?

THE LARGE AND
THE SMALL

In real life, things wear out, and there is an arrow of time. But according to the basic laws of physics developed by Newton and his successors, nature has no inbuilt sense of time. The equations that describe the motion of the Earth in its orbit around the Sun, for example, are time symmetric. They work as well "backwards" as they do "forwards". We can imagine sending a spaceship high above the Earth, out of the plane in which the planets orbit around the Sun, and making a film showing the planets going around the Sun, the moons going around the planets, and all of these bodies rotating on their own axes. If such a film were made, and were then run through a projector backwards, it would still look perfectly natural. The planets and moons would all be proceeding in the opposite direction around their orbits, and spinning in the opposite sense on their axes, but there is nothing in the laws of physics which forbids that. How can this be squared up with the idea of an arrow of time?

Perhaps it is better to pinpoint the puzzle by looking at something closer to home. Think of a tennis player, standing still and bouncing a tennis ball on the ground, repeatedly, with a racket. Once again, if we made a film of this activity and ran it backwards it wouldn't look at all odd. The act of bouncing the ball is reversible, or time symmetric. But now think of the same person lighting a bonfire. He or she might start with a neatly folded newspaper, which is spread into separate sheets which are crumpled up and piled together. Bits of wood are added to the pile, a burning match applied, and the fire takes hold. If *that* scene were filmed and projected backwards, everyone in the audience would immediately know that something was wrong. In the real world, we never see flames working to take smoke and gas out of the air and combine them with ash to make crumpled pieces of paper, which are then carefully smoothed by a human being and neatly folded together. The bonfire-making process is irreversible, it exhibits an asymmetry in time. So where is the difference between this and bouncing a tennis ball?

One important difference is that in the bouncing ball scenario we simply have not waited long enough to see the inevitable effects of increasing entropy. If we wait long enough, after all, the tennis player will die of old age; long before that the tennis ball will wear out (and I am not even considering the biological needs of the tennis player involving food and drink). Even the example of the planets orbiting around the Sun is not really reversible. In a very, very long time (thousands of millions of years) the orbits of the planets will change because of tidal effects. The rotation of the Earth, for example, will get slower while the Moon moves further away from its parent planet. A physicist armed with exquisitely precise measuring instruments could detect these effects from even a relatively short stretch of our film, and deduce the existence of the arrow of time. The arrow is *always* present in the macroscopic world.

But what of the microscopic world? In school, we are taught that the atoms which make up everyday things are like hard little balls, which bounce around and jostle one another in *precise* obedience to Newton's laws. Neither the laws of mechanics nor the laws of electromagnetism have an inbuilt arrow of time. Physicists like to puzzle over these phenomena by thinking about a box filled with gas, because under those conditions atoms behave most clearly like little balls bouncing off one another. When two such spheres, moving in different directions, meet each other and collide, they bounce off in new directions, at new speeds, given by Newton's laws; and if the direction of time is reversed then the reversed collision also obeys Newton's laws. This raises some curious puzzles.

One of the standard ways to demonstrate the second law of thermodynamics is with the (imaginary) aid of a box divided into two halves by a partition. Imagine one half of such a box filled with gas, and the other empty (this is only a "thought" experiment; we don't need to actually carry it out, because our everyday experience tells us what will happen). When the partition is removed, the gas from

the "full" half of the box will spread out to fill the whole
box. The system becomes less ordered, its temperature falls,
and entropy increases. Once the gas is in this state, it never
organises itself so that all of the gas is in one half of the
box once again, so that we could put the partition back
and restore the original situation. That would involve decreasing entropy. On the macroscopic scale, we know that
it is futile to stand by the box, partition in hand, and wait
for an opportunity to trap all of the gas in one end.

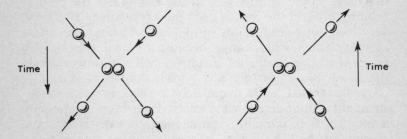

*Figure 1.3/*The way we imagine atoms to collide and
bounce off one another in a gas, obeying Newton's laws
of motion, seems to have no inbuilt arrow of time. The
picture looks equally plausible whichever way we draw
the arrow.

But now look at things on the microscopic scale. The
paths followed by all of the atoms of gas in moving out
from one half of the box – their trajectories – all obey
Newton's laws, and all of the collisions the atoms are
involved in along the way are, in principle, reversible. We
can imagine waving a magic wand, after the box has filled
uniformly with gas, and reversing the motion of every
individual atom. Surely, then, they would all retrace their
trajectories back from whence they came, retreating into
one half of the container? How is it that a combination of
perfectly reversible events on the microscopic scale has
conspired to give an appearance of irreversibility on the
macroscopic scale?

There is another way of looking at this. In the nineteenth century, the French physicist Henri Poincaré
showed that such an "ideal" gas, trapped in a box, from

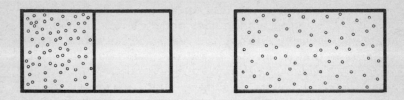

Figure 1.4/ But when we look at the behaviour of a large number of atoms in a box of gas, it is easy to see the asymmetry of time. When a partition in the box is removed, the gas spreads out to fill the whole box. We can easily tell which is "before" and which "after", with no need for labels or arrows on the diagram. The explanation is that Newton's laws do not tell the whole story about collisions between atoms.

the walls of which the atoms bounce with no loss of energy, must eventually pass through every possible state that is consistent with the law of conservation of energy, the first law of thermodynamics. *Every* arrangement of atoms in the box must happen, sooner or later. If we wait long enough, the atoms moving about at random inside the box *must* all end up in one end, or indeed in any other allowed state. Putting it another way, if we wait long enough the whole system must return once again to any starting point.

"Long enough", however, is the key term here. A small box of gas might contain 10^{22} atoms (that is, a 1 followed by 22 zeros), and the time it would take for them to return to any initial state would probably be many, many times the age of the Universe. Typical "Poincaré cycle times", as they are known today, have more zeros in the numbers than there are stars in all the known galaxies of the Universe put together, numbers so big that it doesn't really make any difference whether you are counting in seconds, or hours, or years. Those huge numbers represent the odds against any particular state occurring, by chance, during any particular second, or hour, or year that you happen to be watching the box of gas.

This provides the standard "answer" to the puzzle of how a world that is reversible on the microscopic scale can be irreversible on the macroscopic scale. The irreversibility, the traditionalists allege, is an illusion. The law of increasing entropy is a *statistical* law, they say, in the sense that a decrease in entropy is not so much specifically forbidden as extremely unlikely. If we watch a cup of lukewarm coffee for long enough, according to this interpretation, it will indeed spontaneously produce an ice-cube while the surrounding liquid gets hotter.* It just happens that the time required for this to occur is so much longer than the age of the Universe that we can, for all practical purposes, ignore the possibility.

This interpretation of the law of increasing entropy as a statistical rule, not an absolute law of nature, has recently been questioned, as we shall shortly see. But long before that challenge was raised, the probabilistic interpretation led to one of the strangest theories about the origin of the Universe as we know it, and of the arrow of time – a theory well worth describing for its curiosity value, even though it is no longer taken seriously.

AN IMPROBABLE UNIVERSE

The idea that all states allowed by the first law of thermodynamics are constantly recurring, if we wait long enough, is hard to reconcile with the implication of the second law of thermodynamics, that entropy is increasing and that there is a unique direction in which time's arrow points. Ludwig Boltzmann, who was born in Vienna in 1844 and became one of the great developers of the ideas of thermodynamics, found a way to reconcile the two ideas. But it meant abandoning the "common-sense"

---

* If this sounds vaguely familiar, perhaps you've been reading *The Hitch Hiker's Guide to the Galaxy*, where Doug Adams describes the workings of the "infinite improbability drive". Genuine scientific theorising can indeed be as strange as any fiction.

understanding of the flow of time, and also introducing
the idea of a universe unimaginably more vast than anything we can see.*

Poincaré's work had shown that any closed, dynamic
system must repeat itself indefinitely, given enough time,
passing through every possible new state. This does not
solely apply to boxes of gas, but to *any* system, including
the Universe itself, or our Milky Way Galaxy. In a truly
infinite universe, extending forever in space and with an
eternal lifetime, anything which is not explicitly forbidden
by the laws of physics must happen somewhere, at some-
time (or, indeed, in an infinite number of places, and an
infinite number of times). Boltzmann's argument was that
the entire observable Universe must represent a small, local
region of a much bigger universe, a region in which one
of those very rare, but inevitable, fluctuations, equivalent
to all the atoms in the box of gas gathering in one end, or
the ice-cube forming in a cup of coffee, had happened, on
a grand scale.

In Boltzmann's day, "the Universe" meant our Milky
Way Galaxy. It wasn't until the twentieth century that
astronomers fully appreciated that our Galaxy, containing
trillions of stars, is just one among many millions of
galaxies scattered across a vast sea of space. But that
doesn't affect the argument – it simply adds a few more
zeros to numbers that are already far too large for any
human comprehension.

The argument goes like this. Suppose that there is a
universe out there which is vastly bigger than anything
we can see, but which is, overall, in thermal equilibrium
with maximum entropy. According to Boltzmann:

> In such a universe, which is in thermal equilibrium
> and therefore dead, relatively small regions of the size
> of our galaxy [Universe] will be found here and there;
> regions (which we may call "worlds") which deviate

---

* There is a convention by which Universe, capitalised, refers to the real world of
stars and galaxies that we can see, while universe, without the capital letter,
refers to one of the more or less speculative ideas of the theorists. I shall try to
follow this convention. Similarly "Galaxy" means our Milky Way, while "galaxy"
refers to any old galaxy anywhere in the Universe.

significantly from thermal equilibrium for relatively
short stretches of those "aeons" of time.*

The only change we have to make to bring the description up to date is the insertion of the word "Universe",
What Boltzmann said is simply that we are living in a
bubble of space where there has been a small, local deviation from equilibrium, and which is now returning to
the long-term natural state of the greater universe. As
Boltzmann pointed out, the arrow of time in such a low-
entropy bubble will point from the less probable state to
the more probable state, in the direction of increasing
entropy. There is no unique arrow of time referring to the
whole universe, but only a local arrow of time which
applies to the region we happen to be living in. The bizarre
nature of this interpretation of time's arrow (of course,
Boltzmann didn't use this term, which hadn't been
invented then) can best be seen from a diagram (Figure
1.5), which makes it clear that *everywhere* in the bubble
of low entropy the arrow points towards the high entropy
state.

According to this point of view, the Universe is an extra-
ordinarily improbable and unlikely event, which has inevitably happened in an infinite universe. Boltzmann's own
motivation in putting the idea forward is clear from his
own words:

> It seems to me that this way of looking at things is
> the only one which allows us to understand the
> validity of the second law, and the heat death of each
> individual world, without invoking a unidirectional
> change of the entire universe from a definite initial
> state to a final state.*

But that idea of unidirectional change, which Boltzmann
dismissed, is *exactly* how modern cosmologists view the
Universe! Boltzmann, of course, knew nothing of the big
bang theory or the cosmic background radiation, and lived
at a time when it "went without saying" that the Universe

---

* Both quotes are taken from page 254 of *Order out of Chaos* by Ilya Prigogine
and Isabelle Stengers. They originate in Boltzmann's *Lectures on Gas Theory*,
reprinted by the University of California Press in 1964.

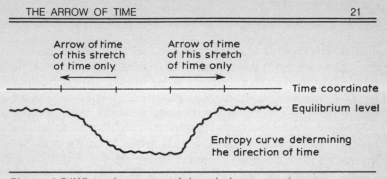

Figure 1.5 / When the arrow of time is interpreted as an indicator of the direction in which entropy increases, it is possible to imagine that the Universe as we know it has been produced by a random fluctuation of entropy. In that case, wherever an observer may be in the region of low entropy, the local "arrow of time" will point in the direction of increasing entropy. Perhaps the arrow of time is not a universal absolute, after all?

did not have a definite origin in time, and would not have a definite end. Today, most cosmologists think differently, and the idea of a universe with a birth, finite lifetime and death is widely accepted, at least as a possibility. Boltzmann's improbable universe is a curiosity of history, one made even less likely as a description of the real world, as we shall shortly see, by the new interpretation of thermodynamics. But his discussion of what we now call the arrow of time raises an interesting point that is well worth going in to in a little detail here, and which should be borne in mind throughout the rest of this book.

There is a fundamental difference between the concept of an arrow of time, pointing in a certain direction, and our subjective impression of a *flow* of time, moving in a certain direction. The point is implicit in Boltzmann's discussion, but it has been made more forcefully in recent years by Paul Davies, of the University of Newcastle upon Tyne. Davies has used the example of a movie film, like our imaginary film of a fire being started. He points out* that if such a film of a time-asymmetric process were made,

---

* For example, in *Space and Time in the Modern Universe*.

and then the film was cut up into its individual frames, and these were jumbled together, it would still be possible to replace them in their correct order by studying the differences between the individual pictures. It isn't necessary for the film to "run", or for time to flow, in order for the inherent asymmetry to be apparent. The flow of time is a psychological phenomenon, which arises from the way we interact with a time-asymmetric Universe.

The analogy Davies uses to help make this clear is with a compass needle on board a ship at sea. The needle always points to the north magnetic pole, indicating an asymmetry. But that does not mean that the ship is always sailing north. The ship could be sailing due south (or in any other direction) and the arrow would still point north. Or, if we wished, we could choose to make our compass needles so that the arrows on them point south. They would still be just as useful for navigation, even though the arbitrary convention assigning a "direction" to them had been reversed. It is natural that we should define the direction on which the arrow of time points as the direction in which we perceive time flow. But it is important to remember that the asymmetry is a built-in feature of the Universe, while our perception of the flow is a phenomenon that nobody can claim to understand. In particular, if time "flowed" backwards it would make no difference to the asymmetry, and all of the thermodynamic arguments would still stand.

It may seem like philosophical hair-splitting, but hold on to the idea; it's going to be important later on. First, though, let's look at those new ideas in thermodynamics, which turn the traditional interpretation on its head, pull the rug from under Poincaré and Boltzmann, and hold that the irreversibility really is a fundamental feature of our Universe, once the arrow of time is defined.

THE IRREVERSIBLE UNIVERSE

The new ideas stem mainly from the work of Ilya Prigogine, who was born in Moscow in 1917 but has been associated with the Free University of Belgium since 1947, and more recently also with the University of Texas, Austin. He received the Nobel Prize for Chemistry in 1977, for his work on non-equilibrium thermodynamics. But his ideas have yet to filter through into the textbooks used by most students of thermodynamics, even at university level.

Prigogine's attack on the problem of reconciling macroscopic irreversibility and microscopic reversibility can be understood in terms of the Poincaré recurrence time, by taking on board some basic ideas from quantum theory. Quantum physics, developed in the first half of the twentieth century, provides a better description of the behaviour of atoms and smaller particles than the older, classical ideas of electromagnetism and Newtonian mechanics. It is only with the aid of quantum physics that modern scientists are able to understand the workings of atoms, and the interactions between particles and electromagnetic fields. We don't need to go into all the details here,* but there are two key features of quantum physics that are relevant to the thermodynamics of the Universe.

The first important point is that the equations of quantum physics, like those of classical physics, are time-symmetric. There is no arrow of time built in to quantum physics, and reactions, or interactions, can, according to the equations, proceed just as happily "backwards" as "forwards". That seems to leave us in the same bind that Boltzmann was stuck in with the conflict between the reversibility of Newtonian mechanics and the wearing out of the real world. But the second salient feature of the new physics gets us off that particular hook.

Werner Heisenberg, a German scientist who made major

---

* If you want the full story of the quantum revolution, see my book *In Search of Schrödinger's Cat*.

contributions to the development of quantum theory, discovered that the equations do not allow us to make a precise measurement of both the position and the momentum of a particle at the same time. We cannot know, as a matter of principle, exactly where a particle is *and* where it is going. We can determine either property *on its own* as accurately as we like, but the more precisely we measure position the less information we have about momentum, and vice versa. The same rule, incidentally, applies to other pairs of what are called "conjugate variables", but that need not concern us for now.

When Heisenberg first reported this uncertainty principle, many people thought that he was talking about some limitation on the practical skills of human observers, and meant that although an electron, say, could be in a definite position and be moving with a definite velocity and momentum, it was forever beyond our skill to measure them both at the same time. Indeed, many people today still think that this is what quantum uncertainty is all about. But they are wrong. The essential feature of Heisenberg's discovery – indeed, in many ways the essential feature of quantum physics – is that the entity we call "an electron" *does not possess* both a well-defined position *and* a well-defined momentum, simultaneously. There is an *intrinsic* uncertainty, which has to do with the way the Universe is put together, and has nothing to do with the skill, or otherwise, of human experimental physicists.

This is not common sense, but why should it be? Our common sense is based on everyday experience with objects on the human scale, and on that scale the uncertainty effect is far too small to notice. We have no basis to know what the "common sense" of things on the scale of atoms and electrons is, except with the aid of theories that predict how collections of such particles will respond in certain circumstances. The theory that makes the best, most accurate and most consistently correct predictions is quantum theory, including the uncertainty principle. Indeed, this is only the tip of the iceberg of quantum oddity, for the best interpretation of the "meaning" of quantum theory is that there is *no* underlying "reality" which builds up to make the macroscopic world. The only reality lies in the actual events we observe – the swing of a needle across a dial when an electric current flows, the click of a geiger counter as a charged particle passes through its detector, and so on. Nothing is real unless it is observed, say the quantum physicists, and there is no point in trying to imagine what atoms and electrons are "doing" when they are not being monitored.

All of these ideas carry over into Prigogine's version of thermodynamics. The reality that we observe is the macroscopic world, with its inbuilt arrow of time and asymmetry. Why, he asks, should we imagine that this world is built up in some way from the behaviour of countless tiny particles obeying precisely reversible, time-symmetric laws of behaviour? To Prigogine, the macroscopically derived second law of thermodynamics is the fundamental truth, a *precise* law that always holds, not a statistical rule of thumb that applies, more or less, for most of the time. It is the apparently time-symmetric behaviour of little spheres bouncing off one another that he regards as an approximation to reality. "Irreversibility," he says, "is either true on *all* levels or none. It cannot emerge as if by a miracle, by going from one level to another." *

We can see what he is driving at, and the direct relevance of quantum physics to thermodynamics, by looking at another example of a closed system, the kind which Poincaré said ought to return to its initial conditions, given enough time. We start, once again, with a box full of gas, but this time make it a little bit more complicated, by placing in the box a smoothly sloping hill of material, completely symmetrical, rising up to a rounded top. Imagine a perfectly round ball balanced precisely on top of that hill, with the box shut, as usual, to keep it thermodynamically isolated from the rest of the Universe. What will happen to the ball? Obviously, it will roll off the top of the hill. But which way will it roll? The direction taken by the ball, and the subsequent history of the material in the box, will depend on some tiny accumulation of little nudges by the atoms of gas bouncing off it. There will be a minuscule pressure pushing the ball one way, just by chance, and off it will roll.

* *Order out of Chaos*, page 285.

According to Poincaré, eventually the ball will return to
its starting place. When the ball rolls off the hill, it gives
up energy to the gas, energy derived from the fall of the
ball, ultimately from gravity. If we wait long enough
(many, many times the age of the Universe!) it will just
happen that most of the atoms of gas bouncing off the ball
will be moving in the same direction, and will give it a
little push, with precisely the same amount of energy that
it previously gave up to the gas, so that it rolls back up
the hill while the gas cools down. There will be other
occasions when the ball receives a push in the wrong
direction, or one that is too strong, or too weak, to leave it
balanced once again on top of the hill. But after a suitably
long interval, there will be an occasion when the push is
exactly sufficient to return the ball to the top of the hill,
and leave it balanced there. The system has returned, as
predicted, to its original state. Or has it?

If there is the tiniest difference in the way the atoms of
gas now strike the ball, compared with the first time it
was on top of the hill, it will roll off in a different direction,
and the future history of the little world inside the box will
be completely different. And there *must* be tiny differences
in the way the atoms are striking the ball, because
quantum uncertainty makes it impossible to define *any* set
of conditions precisely for the atoms. Even in this very
simple case, we can imagine the ball so precisely balanced
that as tiny a change in the conditions as you like will
alter its future behaviour. The real Universe is vastly more
complicated than this, and complex systems involving
many particles are known to be prone to very strong in-
stabilities, so that a tiny change in starting conditions
produces a drastic alteration in the system's future behaviour.

Or, if you prefer, think of things in terms of the actual
reversibility of atoms moving in a box of gas. When we
think about the system where the gas in half of the box
spreads out to fill the whole box, it is easy to say "imagine
reversing the motion of every atom simultaneously". The
image this conjures up is of something like a pool table,
with balls moving on it, which suddenly reverse their
trajectories and return to their starting positions. We can't

actually do the trick, but we can, indeed, imagine it. But think what the simple statement really means. It requires that the position of *every* atom should be precisely determined, and that its velocity should *simultaneously* be precisely determined, and then exactly reversed while the atom stays in precisely the same position. But quantum physics tells us that this is impossible! No atom does have the two characteristics (precise position) and (precise velocity) at the same time! The laws of nature, as they are best understood today, make it impossible to reverse the direction of every atom in the gas, *in principle*, not just because of the practical limitations set by human skills.* There is no magic powerful enough to do the job. So, once again, we find that a system that seems to be reversible is, in fact, irreversible.

This is just a very simple interpretation of one aspect of Prigogine's reinterpretation of thermodynamics. But the gist of his message is plain, and important. *No matter how long* we sit and watch a lukewarm cup of coffee, it will *never* spontaneously give birth to an ice-cube and heat up; no matter how long we sit by a box of gas, it will *never* all congregate in one half of the box so that we can trap it in a state of lower entropy. The second law of thermodynamics is an absolute ruler of the Universe.

These are difficult ideas, made no easier to absorb by the tie-in with the quantum theory idea that there is no underlying reality to nature. To Prigogine, reality lies only in the irreversible processes going on in the world – not in the "being", as he puts it, but in the "becoming". I warned you that this would be a book raising questions, rather than giving answers, about the nature of the Universe. Is Prigogine's version of thermodynamics better than the traditional, standard version? As of now, it is largely a matter of personal prejudice which version you prefer; but Prigogine has one powerful argument in his favour, sufficient to persuade me to side with him until he is proved

---

* Even atoms are relatively large by the standards of quantum physics, so the effect of quantum uncertainty is still small at this level. But the same argument applies to the deeper level of electrons and protons, so the argument is completely valid. And, of course, even the tiniest deviation from perfection is still imperfection.

wrong – nobody has ever seen a violation of the second law of thermodynamics, and until they do it seems best to accept it as just that, an unbreakable *law* of nature.

For the story of the Universe at large, though, that is as far as we need take the debate for now. We *do* live in a region of increasing entropy, an expanding bubble of dark space dotted with a few bright lights, the stars and galaxies. Time *does* flow, as we perceive it, forwards from the big bang to the omega point, the death of the Universe. The ultimate fate of the Universe, it turns out, depends on just how much matter it contains, not only in those bright stars but in dark forms between the stars and galaxies. But it also depends on how the Universe got to be the way it is today – and the present "best buy" among the cosmological theories is a far cry indeed from Boltzmann's vision of an enormous Poincaré cycle.

CHAPTER TWO

THE UNIVERSE IN A NUTSHELL

To know the future we must understand the past. To understand the nature of the Universe we live in, and to gain some insight into its likely fate, we must understand, as best we can, where it came from and how it came to be as we see it today. Cosmologists, in fact, talk far more readily about the birth of the Universe than they do about its ultimate fate. For, although nobody is sure quite how the Universe will end, there is a general consensus among astronomers that the Universe started out in a hot, dense state and has been expanding ever since – for, roughly speaking, 15 billion years.

The idea of the big bang is so well known that few people stop to think just how dramatic an achievement of the human mind the theory is. It is one of the most remarkable features, not just of twentieth-century science but of all human thought, that in the 1980s astronomers are able to describe in detail an essentially complete, self-consistent theory, or model, of how the Universe was born in a fireball of incandescent radiation some 15,000 million years ago, and how that expanding fireball developed into a cold, dark Universe of space dotted with islands of light and life, the galaxies of stars which include our own Milky Way. The

wisest cosmologists are always wary of claiming that they
have in fact found some ultimate truth. They are careful
to refer to their ideas about the big bang only as the
"standard model", and to emphasise that the impossibility
of observing the totality of the Universe means that even
their best ideas can never be put on so secure a footing as,
say, Newton's law of gravitation as applied to the behaviour of falling objects here on Earth. They are right to
express these reservations, but in everyday terms the
standard model is as well established as it possibly can be,
and is the best and most reliable indication of how
everything got to be the way it is that humankind has
ever had. In this book, and especially in this chapter, I
shall be dealing with the standard model, which, my
cosmologist friends would stress, is only a model and may
not hold the ultimate truth. But I shall describe the model
as if it does provide a complete description of the real Universe, and I shall use the term "Universe" when I should,
strictly speaking, say "model". It *is* the best description of
the Universe, so a little literary licence should not upset
the cosmologists too much.*

THE SCALE OF
THE UNIVERSE

The first thing to grasp about the Universe is its size. The
Universe is big – vastly bigger, in both space and time,
than anything within human experience. Distances across
the Galaxy can be conveniently measured in terms of the
time it takes light to traverse them. Light travels through
space at a constant 30,000 million centimetres every
second, rather more than 1,000 million kilometres an

---

* Inevitably, this chapter in particular covers ground that may be familiar to
some readers, especially those who have read my own earlier books. If you think
you already know about the cosmologists' standard model of the big bang, feel
free to move on to Chapter Three, or to regard this as a quick refresher course in
cosmology.

hour. At this speed, it still takes more than four years to travel across space to us from the nearest star. That star, Alpha Centauri, is therefore said to be 4.3 light years away – a light year is a measure of distance, the distance travelled by light in one year, not of time. On that measure, our Milky Way Galaxy is a flattened disc of stars about 100,000 light years across and 2,000 light years thick, embedded in a halo of stars stretching out across a sphere 500,000 light years in diameter (and there may be more to the Galaxy than the bright stars, as we shall discover).

By measuring the brightness of individual stars in the nearest galaxies, astronomers have been able to work out their distances. The Andromeda galaxy, also known as M31, can just be seen, using the unaided human eye, as a faint patch of light in the constellation Andromeda. It is the most distant object visible to the naked eye, a little more than two million light years away from us. We see it now by light which left two million years ago, when our ancestors were already of the genus *Homo*, but still two steps short of becoming *Homo sapiens*.

Yet to the cosmologists the Andromeda galaxy is such a near neighbour that its behaviour tells us nothing about how the Universe got to be the way it is. To probe those mysteries, they have to look further away across space, which, because of the time taken by light to reach us, in a sense implies looking further back in time. The investigations extend to galaxies scores of millions of light years away, and the key discovery thrown up by these investigations is that each group of galaxies in the Universe is moving away from every other group – the Universe as a whole is expanding.

The discovery was made in the 1920s by the American astronomer Edwin Hubble. Hubble and his colleagues found that the expansion of the Universe follows a simple law – the velocity with which two groups of galaxies recede from one another is proportional to the distance between them. This is known as Hubble's law, and is usually stated, from our terrestrial viewpoint, in the form that the velocity of recession of a distant galaxy *from us* is proportional to its distance from us. But, in fact, it is a universal law that applies to any pair of galaxies that are not in the same group, or cluster, as each other – there is nothing special about our location in the expanding Universe.

Hubble's discovery was based upon a chain of measurements, stepping-stones out into the Universe. The trick of measuring the brightnesses of individual stars in other galaxies only works for our nearer neighbours in the Universe, where individual stars can be distinguished (resolved) with the aid of large telescopes. It works especially well for certain types of star, called Cepheid variables, which vary in brightness in a regular way which enables astronomers to determine their true brightness. The technique is rather like measuring the distance to a hundred watt lightbulb by measuring its apparent brightness and working out how much its intrinsic brightness has been reduced by distance; but it was enough to enable astronomers to take the first steps beyond the Milky Way.

To go further, they needed other tricks, using estimates of the brightness of whole clusters of stars, and then of whole galaxies themselves. These were just sufficient to enable Hubble to propose his famous redshift-distance relation.

When any object is moving away from us, light emitted by the object is stretched on its way to us. This makes the wavelength of the light longer, and in terms of the spectrum of visible light that means moving the wavelength towards the red end of the spectrum. In a similar way, an object moving towards us emits light waves which are squeezed closer together by the motion, shifted to shorter wavelengths – a blue shift. But, of course, the effect is only noticeable at all if the objects are moving at a respectable fraction of the speed of light – although there is, in principle, a tiny redshift or blueshift in the light from everyday objects moving about here on Earth, it is far too small for our eyes to notice, even if the moving object we are looking at is Concorde. Sound, though, is different. The speed of sound is much less than the speed of light, only a few hundred kilometres per hour, and everyday objects do move at a respectable fraction of the speed of sound. The result is that our ears *can* detect the equivalent of a redshift or a blueshift in noise made by moving objects. It is called the Doppler shift, and explains, for example, why the note of the whistle of a train sounds deeper if the train is moving away from us.

Hubble and his colleagues found that for all except our nearest neighbours the light from other galaxies is redshifted, which means that they are moving away from us (and, of course, from each other). As far as estimates of distance can establish, the redshift of every group of galaxies is proportional to its distance from us. Since the redshift is proportional to the velocity of recession, this is the basis of Hubble's Law, that recession velocity is proportional to distance, the cornerstone of modern cosmology.

THE EXPANDING UNIVERSE

We do not live in a special place in the Universe, even though from our point of view all the distant galaxies are receding uniformly, as if rushing away from us. Hubble's law turns out to be the only kind of redshift-distance relation that would look exactly the same from whichever galaxy in the Universe happened to be home. Imagine a balloon, covered in spots of paint, which is being slowly inflated. Every spot of paint moves away from every other spot of paint. And, if you carry through the mathematics, it turns out that an imaginary "observer" sitting on any of those paint spots will see the others moving away in line with Hubble's law – more distant spots recede faster. If one paint spot starts out, relative to the observer, twice as far away as another, then by the time the distance to the nearer spot has doubled so has the distance to the further spot, which seems to have moved twice as fast. Whichever paint spot you view from, the story is the same; in the real Universe, whichever galaxy you view from the story is the same.

The galaxies do not move "through space", any more
than the paint spots move through the skin of the balloon.
Like the balloon, empty space (which Einstein taught us is
a tangible thing which obeys the laws embodied in the
equations of general relativity) is expanding, and carrying
the galaxies along for the ride. There is no centre to the
expansion in space, but the expansion does imply that
there must have been a beginning to the Universe in time
– a moment of creation.

If the galaxies are moving further apart all the time, as
empty space expands, then long ago they must have been
packed more closely together. When they were closer together, the Universe must have been hotter, for the same
good, thermodynamic reasons that a bicycle-pump gets
hotter when you use it to squeeze gas into the bicycle tyre.
Go back far enough, and the galaxies must have been
squeezed into an amorphous lump, one giant ball of star
stuff as hot as the Sun is today. And before that? The established cosmological wisdom has it that this analogy can be
pushed right back into the past to a time when everything
in the Universe was packed into a superhot, superdense
state, before the big bang. It is by measuring how fast the
galaxies are moving apart today, and running their calculations backwards in time, that cosmologists can estimate
the time that has elapsed since the big bang as roughly 15
billion years. Some estimates give a figure as low as 10
billion years, some as high as 20 billion years. The reason
for the differences in the estimates are important, and I
will come on to them later. But the expansion of the Universe tells us that there *was* a beginning, and that is the
more important point for now.

THE BIG BANG

Today, some physicists believe that they can explain how
the Universe appeared in a superhot, superdense state at
the moment of creation itself. These ideas are relevant to
the discussion of the ultimate fate of the Universe, and will

be dealt with in the next chapter. But the standard big bang model, which dates from the mid-1960s and therefore counts as a very old and respectable feature of cosmological theory (remember, the discovery of the expansion of the Universe only dates back sixty years) deals with events from a fraction of second after the moment of creation, the beginning of time when, if the expansion is taken at face value, the Universe appeared with infinite density at an infinite temperature, occupying a mathematical point of spacetime.

Within a ten-thousandth of a second after the moment of creation, the temperature of the Universe had cooled to a mere 10^{12}K (a thousand billion K), the density was roughly the same as the density of the nucleus of an atom today, and the conditions were such that they can be described in terms of the everyday laws of physics used routinely by scientists who work with atomic nuclei and elementary particles. The standard big bang "only" takes us back as far as one ten-thousandth (10^{-4}) of a second after the moment of creation, but it can describe in beautiful detail everything that has happened since then.

There are some mysteries about what went on in that first split second, although even these are being cleared up by the latest cosmological models. The greatest of these mysteries is that the Universe contains any matter at all. At such very high temperature as existed in the first ten-thousandth of a second, the stable form of energy is in radiation, as highly energetic photons. In thermodynamic equilibrium, a sea of hot photons can constantly produce particles like those of everyday matter, in line with Einstein's equation $E = mc^2$. A photon with energy E can turn into a pair of particles which together have a mass m given by the equation, and those particles can in turn combine to produce another burst of radiation with energy E. Because c is the speed of light, and c^2 is a very big number, you need an enormous amount of energy to make even a modest particle, like an electron. This is the same as saying that the radiation must be very hot, which is why the process is rare today. But in the early split seconds of the life of the Universe, photons and particles were interchangeable. It was only as the Universe cooled that the reactions became impossible, and the present quantity of matter "froze out" as the Universe expanded.

How did this happen? I said that a photon can be converted into a pair of particles, and vice versa. But not just any old pair of particles, quite apart from the need to balance energy and mass. I should have said, a photon can be converted into a particle and its antiparticle partner. For each of the types of particle of ordinary matter, such as the electron, has a mirror image counterpart – in this case, the anti-electron, or positron. Antiparticles are in many ways the opposites of their partners, and the electric charge on a positron, for example, is the same in magnitude as that on the electron, but positive instead of negative – hence its name. When a particle and its antiparticle counterpart meet, they annihilate in a burst of radiation, following $E = mc^2$.

The same rules apply to particles more massive than electrons. The important ones in the world today are neutrons and protons, which are both members of a family termed baryons, and which are the stuff of which atomic nuclei are made. Electrons, protons and neutrons are the stuff of the material world, making up the bulk of all the stars visible in all the bright galaxies in the Universe. But why should there be any at all?

As I have described it, the flow of reactions by which radiation turns into particle-antiparticle pairs, and back into radiation, is symmetrical. The Universe, we think, was born as a fireball of radiation, and when it was young and hot that radiation could produce neutrons, protons and electrons, each with their antiparticle partners. Once the Universe had cooled to the point where no more baryons could be made in this way (which it had done by one ten-thousandth of a second after the moment of creation), why didn't all the baryons and antibaryons annihilate one another to leave nothing but cooling radiation to fill the expanding Universe? They very nearly did – but not quite.

The problem of why this should be so has puzzled physicists and cosmologists for years. The latest theories of physics, however, suggest that there is a tiny imbalance built into the laws of nature, an imbalance which favoured the production of a few more baryons than antibaryons, plus precisely the same excess of electrons over positrons, in the first instants following the moment of creation. In the fireball, for every billion and one baryons, there were only a billion antibaryons; for every billion and one electrons, just a billion positrons. Such a small imbalance hardly mattered, until the Universe cooled and no more particles were being made out of hot photons. Then, it became vital.

These theories are as yet only approximate guides, with many details to be resolved. But the fact is that the Universe *did* emerge from those earliest instants with a tiny surplus of matter over antimatter. Just how small that surplus was can be seen by comparing the number of baryons (protons plus neutrons) estimated to be in all the stars of all the visible galaxies with the number of photons of the cosmic background radiation, the redshifted remnant of the big bang fireball, filling all the empty space between the visible galaxies. In round terms, there must be a billion photons for every baryon in the visible Universe, judging by the measured temperature of the present-day background radiation. And that is a measure of the tiny size of the irregularity in the laws of physics, a one in a billion discrepancy, that permits us to be here to wonder about the origin of the Universe itself.

The precise number, a billion photons for every baryon, itself turns out to be important in the processes which went on as the fireball continued to cool in the seconds and minutes following the first ten-thousandth of a second. This may have a bearing on the ultimate fate of the Universe, but the details will have to come in their place. In this thumbnail sketch of the big bang, there is only space for the briefest outline of those processes.

Once that tiny trace of matter, in the form of neutrons and protons, had settled out of the cooling Universe, leaving the radiation to go on its way and ultimately to be detected by us as a weak hiss of radio noise with a temperature of just under 3 K, nuclear physics could begin. All atoms are composed of relatively tiny, very dense cores, called nuclei, which are composed of protons and neutrons. Each proton carries one unit of positive charge; neutrons, as their name suggests, are electrically neutral. The number of positively charged protons in each atomic nucleus is exactly balanced by the number of negatively charged electrons associated with the atom, forming a cloud much bigger than the nucleus itself. The single particle called the proton forms the nucleus of an atom of the lightest element, hydrogen. In a hydrogen atom, one proton is associated with one electron. There is also another form of hydrogen, heavy hydrogen or deuterium, in which one proton and one neutron are together in the nucleus, still associated with one lonely electron.* Moving up the scale of complexity, the next lightest element is helium, and every atom of helium has two protons in its nucleus and two electrons outside the nucleus. One variety of helium, called helium-3, also has a single neutron in its nucleus (which therefore contains three particles, hence the name); another variety, helium-4, has a nucleus made up of two protons plus two neutrons.

So the complexity of chemistry builds up, through elements such as oxygen, with eight protons in each nucleus, and iron, with 26 protons, to uranium, one of the most massive naturally occurring atoms, with 92 protons. The more protons a nucleus has, the more neutrons it can hold on to as well – the most common form of iron has 30, giving 56 baryons in all, while one type of uranium has no less than 143. But these heavy elements have no place in the story of the big bang. They are made inside stars, by thermonuclear reactions in which more complex nuclei are built up from the simplicity of hydrogen and helium.

Hydrogen, obviously, originates in the big bang. As long as you have protons and electrons, once the Universe is cool enough for them to be able to hold together electrically without being knocked apart by energetic photons, you will inevitably have hydrogen atoms. Slightly less obviously, helium also comes out of the big bang. You need energy to make two protons stick together, because they are repelled by their positive electric charges. Once two protons can be pushed close enough together, however (in

---

* The reasons why atoms are made up like this are discussed in my book *In Search of Schrödinger's Cat.*

effect, once they are touching), they come under the influence of another force, called the strong nuclear force. This attracts any proton or neutron to any other proton or neutron. It overwhelms the electric repulsion, but only once the two have been pushed together. In effect, this happens if a brew of neutrons, protons and radiation is hot enough. For the particles, more heat means that they move faster, carrying more kinetic energy. So, when two protons collide with one another there is a good chance that they will smash right into each other, close enough for the strong force to get to work, instead of being slowed and bounced apart, without ever touching, by electric repulsion.

Conditions in the later stages of the big bang were just right to push protons together and enable many of them to stick to each other in pairs in this way, also capturing neutrons and forming nuclei of helium. A tiny trace of even more complex nuclei, notably of lithium-7, was also produced, but the cooling of the Universe proceeded so rapidly that by the time its composition had been converted into 75 per cent hydrogen and 25 per cent helium it was already too cold for the trick to work any more. This happened within four minutes of the moment of creation, as the temperature fell below 900 million K. Only neutrons safe inside helium nuclei survived this stage in the evolution of the Universe, because an isolated neutron is unstable, and will survive for only a few minutes on its own before ejecting an electron and turning into a proton.

The precise amount of helium formed in the big bang, and the trace of other elements such as lithium-7, depend, among other things, on the temperature of the fireball, the rate at which the Universe is expanding (and therefore cooling) and the exact balance between the roughly one in a billion baryons and the mass of hot photons that still dominated the evolution of the Universe then. These factors also play a part in determining the future evolution of the Universe, and one way of getting a grip on how the Universe will change as it ages is by studying old stars to see exactly what proportion of helium they contain, and plugging the numbers back in to the equations of the standard big bang. But before we move on to that, there are some puzzles about the real Universe to be resolved, puzzles which cannot easily be explained by the standard model, but which fall within the province of the latest ideas about the moment of creation itself, and which give a broad hint to the ultimate fate of the Universe.

SOME COSMOLOGICAL PUZZLES

It took cosmologists a long time to appreciate fully that the Universe we see around us is in a very peculiar state and that the standard model raises as many questions as it answers. The first peculiarity should be obvious, in the light of Chapter One of this book. Even allowing for the fact that most of the entropy in the Universe is in the form of those photons of the background radiation, at first sight it seems to fly in the face of the second law of thermodynamics to claim that an initially homogeneous, chaotic universe, uniformly filled with matter and radiation, should evolve in a direction that produces more order, in the form of stars and galaxies.

The existence of the background radiation also raises with full force a puzzle that is already apparent from the pattern of galaxies on the sky. The Universe is the same in all directions that we look. Of course, there are variations; the galaxies are not distributed perfectly uniformly. But they are distributed with a high degree of uniformity, and the background radiation is dramatically more uniform. We measure this uniformity in terms of temperature, deviations from the average 2.7 K that typifies the cosmic background. And these deviations are tiny. Wherever we look on the sky, once we make allowances for the Doppler effect of the Earth's own motion around the Sun, and the motion of the Solar System round our Galaxy, the temperature is the same. Why is the Universe so smooth? One way of highlighting the puzzle is to appreciate that when we look in opposite directions in space, we are "seeing" (with our radio telescopes) the background radiation that was emitted in the late stages of the cosmic fireball, certainly less than a billion years after the moment of creation. The radiation from one direction has been travelling all that time across expanding space to reach us; the radiation from the opposite side of the sky has been equally long on its journey. Photons from opposite sides of the sky, that have had no contact with each other and have been journeying for ten billion years or more, have precisely the same temperature!

The Universe, it seems, is a very smoothly structured place. Then again, since the Universe, typified by the background radiation, *is* so smooth, how come there are any irregularities – any galaxies – in it at all? How can galaxies grow in a smooth, expanding universe? These are very real puzzles that the standard model could not answer in the 1960s and 1970s.

But the most remarkable feature of the expanding Universe, once we know that it *is* expanding, is the rate at which it expands. We can imagine, and describe mathematically, two extreme kinds of expanding universe, each defined with precision by the equations of general relativity. The expansion is dominated by gravity, the gravity of all the matter in the Universe holding it together. It is this gravity that will decide the ultimate fate of the Universe. Although it is important to remember that we are dealing with stretching spacetime, not with matter moving through space, the two possibilities are rather like the two possible fates of a rocket fired upwards from the Earth. If the rocket is fired fast enough, it will escape from the Earth and keep going forever. If it has less than the so-called escape velocity, however, it will fall back to the ground.

This translates into general relativistic terms, for the whole Universe, as a choice between being "open" and "closed". An open universe is one which expands forever, with clusters of galaxies always getting further and further apart. A "closed" universe is one which must inevitably fall back on itself, as gravity first halts and then reverses the expansion. On the borderline, poised between these two states, there is the possibility of a so-called "flat" universe, which *just* expands forever, balanced on a gravitational knife-edge.

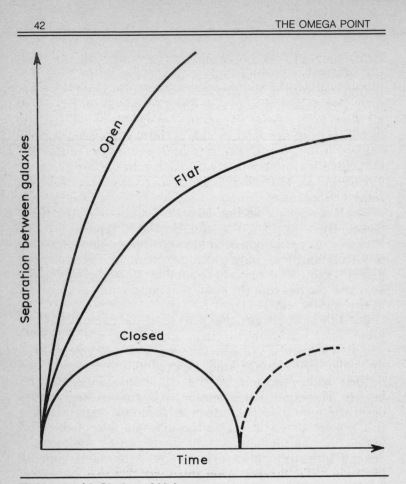

Figure 2.1 / A Choice of Universes
The Universe we live in must conform to one of three
possible types. If it contains enough matter, it is closed
and will one day collapse back on itself. If it contains
too little matter, gravity can never halt its present
expansion, and it is open. Exquisitely balanced on the
dividing line between these two families lie the "flat"
models. Observations show that our Universe is very
nearly flat – but they cannot say which side of the
dividing line it lies.

Astronomers can get a good idea of how rapidly the
Universe is expanding today, by measuring redshifts. And

they can get a rough idea of how much matter there is in the Universe, by counting the numbers of galaxies. There is a critical density of matter in the Universe which corresponds to the flat state. A little more density, and the Universe is closed; a little less and it is open. Of course, the density is always changing, decreasing as the Universe expands and matter thins out. But it happens, logically enough, that if the density of matter in the Universe is enough to make it closed at one stage in its evolution, then it is *always* enough to make it closed, because of the way the thinning and the gravitational slowing down are balanced as the Universe expands. But, as we are about to see, there is still scope for some dramatic changes in the balance between the two, even if the density must always stay one side or other of the critical dividing line.

By measuring the density of the Universe today, and comparing it with the expansion rate, we can determine once and for all whether our Universe is open or closed. Unfortunately, the density of matter in the Universe is hard to estimate. Cosmologists label the density by the letter omega, Ω. The critical density, corresponding to a flat universe, is defined as $\Omega = 1$. And the astonishing feature of our Universe is that the density today lies very close to the critical density, between 0.1 and 2.0 in these units.

Maybe that doesn't sound so remarkable, but think. In principle, the Universe could have had *any* density. A billionth, or a billion times, the critical density, or anything in between, or even anything outside that range. Why should it lie so close to the *only* special density that comes into the cosmological equations of relativity? Surely this cannot be a coincidence – and surely, some cosmologists have speculated, if this is *not* a coincidence then the density of the Universe must be precisely the critical density?

The coincidence is even more spectacular when placed in the context of the standard model. The expansion of the Universe in that model acts to increase the difference between the density of the Universe and the critical value, whatever side of the dividing line it starts out. Although a closed universe is always closed, and an open universe is

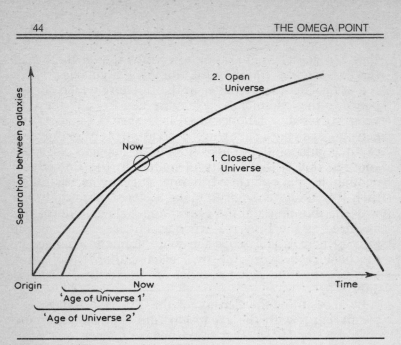

Figure 2.2/The Age of the Universe
The state of the Universe today would look much the
same whether it were just open or just closed. But
because the expansion rate slows down more quickly in
a closed model, the true age of the Universe is much
less, if it is closed, than indicated by simple
measurements of the way galaxies are moving apart.
The simple measurements give an "age" of about 20
billion years; the true age is probably about two thirds
of that figure.

always open, in a sense they get "more closed" and "more
open", respectively, as time passes. The balance point,
$\Omega = 1$, really is an exquisitely fine knife-edge on this picture. The further back in time we look, the closer the density of the Universe must have been to the critical value;
in order to explain the observed state of the Universe today,
with Ω somewhere in the range from 0.1 to 2.0, its density
one second after the moment of creation must have been
equal to the critical density to within one part in 10^{15}
one in a million billion. At earlier times, still crucially important to the big bang explanation of how the Universe
came into being, the Universe must have been even more

"flat", in the cosmological sense of the word. It strains credulity too far to think that this could possibly be a coincidence. Something must have happened to the Universe, even before the end of the first second, to make it so flat. The time has come to see how this dramatic feature of the Universe can indeed be explained, to find out how the new theories that explain the flatness of the Universe raise even more intriguing questions about the relationship between time and the Universe, and to look at the implications of all this for our understanding of where the Universe is going.

CHAPTER THREE

TIME AND
THE UNIVERSE

The uniformity and flatness of our Universe today can best be explained by a new concept, developed in the 1980s, called "inflation". The physics which underpins the new cosmological theories comes not from astronomy, but from particle physics, where researchers have been studying the interactions of so-called "elementary" particles – protons, neutrons, electrons and others – at very high energies. This involves using machines, like those of CERN, the European centre for particle physics research, to smash particles such as protons into one another head on, mimicking, in a modest way, the interactions that occurred in the big bang itself.

THE FORCES
OF NATURE

The Holy Grail of this research, the ultimate truth everyone involved with it is eager to find, is a unified theory which will explain all of the laws of nature in one set of equations. In the macroscopic world, there are just two forces which are important. These are gravity and the electromagnetic force. We all experience gravity, and literally have a feel for what it is: it holds stars together, makes things fall down here on Earth, and determines how the Universe will evolve. The electromagnetic force has two facets, electricity and magnetism, which we all know of, even if our feel for them is slightly more metaphorical. Electromagnetism is the underlying force which determines how light and radio waves, and other forms of electromagnetic radiation, are produced and propagate; it governs the behaviour of electrons within atoms (best understood in terms of its modern form, quantum electrodynamics, or QED) and the way atoms join together to make molecules. Electromagnetism is the crucial force in chemistry, and therefore in life, since our bodies and other living things are made of complex biochemical molecules.

The other two forces of nature operate at more subtle levels, affecting the behaviour of particles such as neutrons and protons, the way atomic nuclei stick together, and the way an isolated neutron will decay into a proton plus an electron. They are called the strong and weak nuclear forces. Today, physicists have a very successful model of the particle world which portrays the proton and neutron as themselves made up of smaller units, called quarks. On that picture, the strong nuclear force is regarded as a manifestation, at the nuclear level, of a more basic interaction involving the quarks. But for simplicity we can still think of the four forces of nature as gravity, electromagnetism, and the strong and weak forces.

The most successful descriptions of how these forces got to be the way they are, and how they affect particles today, see them all as different aspects of some single, unified force, in a similar fashion to the way electricity and magnetism are two aspects of electromagnetism. The collider experiments show that at high enough energies there is no distinction between electromagnetism and the weak force, and that many particle interactions are then best described in terms of a combined force, the "electroweak" interaction. The supposition is that at still higher energies there will be a merging between this electroweak force and the strong nuclear force into one unified interaction, while ultimately, under conditions of very high energy indeed, gravity itself, the weakest of the four forces, would be brought into the compass of a single completely unified theory.

SMOOTHING THE UNIVERSE

All this is relevant to the story of the very early Universe, because, of course, the further back in time we go (in our imaginations) towards the moment of creation the higher the temperature of the Universe, and therefore the greater the amount of energy in a constant volume. The most successful unified theories are the ones which predict that the Universe should indeed have emerged from this superhot fireball with a tiny excess of matter over anti-matter. Now, more speculative, but respectable, refinements of those theories offer an explanation of the uniformity of the Universe.

Protons, neutrons and electrons were not the only particles being created and destroyed in the energetic fireball. As far as we can tell from modern studies of particle interactions, there is one layer of structure within each baryon. The most fundamental units of the particle world are called quarks. There are six different types of quark, which can group together in threes, to make baryons, or in pairs, to make another family of particles called mesons, but can never occur in isolation. Different combinations of quarks can build up a whole variety of more complex particles such as protons and neutrons.* Most of these are unstable, and fall apart into showers of other particles (ultimately, into neutrons and protons) if

---

* Electrons *are* thought to be stable, truly "fundamental" particles. They are part of a family called leptons; leptons and quarks together are the most basic building-blocks of nature yet identified.

they are made in the big colliders today. But in the very early Universe, with lots of energy about, all kinds of exotic particles were being made by pair production and annihilating each other again every split second.

The details need not worry us here. But in this maelstrom of activity, the conditions were constantly changing as the Universe expanded and cooled. As the temperature dropped, it became impossible for the most massive particles to form at all, because there wasn't enough energy left in the photons, so successive types of particle vanished from the scene as time passed. But the changes were not, according to the latest ideas, smooth. Instead, there must have been what the particle physicists call "phase transitions", dramatic shifts from one state, stable at a high temperature, to another state, stable at a lower temperature. The analogy they make is with the three states of water – a gas phase, a liquid phase, and a solid phase. When steam cools, it does so smoothly down to a certain temperature, then stays at that critical temperature while it condenses into water. As it does so, it gives out heat, called latent heat; it is the latent heat given out in this way, rather than the temperature of the steam itself, which makes scalds so severe if steam condenses on human flesh. As the water is cooled further, it reaches another critical temperature where, in spite of the cooling, it stays at the same temperature while turning into ice, giving out more latent heat. These are phase transitions, during which energy is given out as the water molecules settle into lower energy states with higher entropy.

Something similar must have happened in the very early Universe, at a time less than 10^{-33} of a second after the moment of creation, at vastly greater energies. Different theorists have come up with different versions of this kind of event, but they all agree on one thing. The phase transition (or transitions) in the very early Universe would pour energy out, like latent heat being given up, and this would produce a brief, but dramatic, acceleration of the universal expansion. They call this sudden burst of super-expansion "inflation".

It is hard to get a handle on all this in everyday terms, when the timescales involved are so small. *Everything* is happening within less than 10^{-33} sec. That is, a decimal point followed by 33 zeros and a "1". But the expansion that goes on during the split-second of inflation is exponential, with any affected region of the universe doubling in size every tiny split second. Exponential inflation can very soon run away with itself. If one doubling takes 10^{-35} sec, then in the space of just 10^{-34} sec there have been ten doublings, and it takes only 10^{-33} sec for there to be 100 doublings. A hundred doublings are enough to explain the uniformity of the Universe. Starting out with a tiny seed just 10^{-25} of a centimetre across (for comparison, a typical atom is 10^{17} times bigger than this, about 10^{-8} cm across*), that much inflation will blow it up to 10 cm across, the size of a grapefruit. And all within an unimaginably short fraction of a second.

That is all you need to set the early Universe going in accordance with the standard model. After the inflationary "era" ends (all 10^{-33} sec of it!), the Universe continues to expand in line with the standard model, at a much more sedate rate. Any chosen region of the Universe today will take roughly 60 billion years (four times its present age) to double in size again, assuming the expansion continues. But the steadily expanding Universe was created in a very smooth, uniform state, back when it was the size of a grapefruit, by the inflation. The tiny seed from which our entire observable Universe has grown was simply too small to contain any irregularities – it was, indeed, the size of one "grain" in the structure of the pre-inflation state, according to the rules of quantum physics.

What happened to all the other grains in the pre-existing universe? One of the most astonishing implications of the inflation scenario is that our Universe may be surrounded, in some sense, by an array of other universes, regions of spacetime that inflated in slightly different ways, and which we can never gain any knowledge of. But they can have

---

* If it helps, you may like to think of an atom as being one hundred million billion times bigger than the seed from which, with the aid of inflation, our entire Universe has grown. None of these numbers mean much to me; the only thing to grasp is that a region far smaller than an atom inflates, very rapidly (instantaneously, by human standards), to the size of an everyday object, like a softball.

no influence on the fate of our Universe, either, and so will not come in to the rest of the discussion in this book. For now, what matters is that it is inflation that takes the Universe from the microscopic to the macroscopic, smoothing it in the process. And it also makes spacetime flat as it does so.

FLATTENING THE UNIVERSE

The three possible kinds of universe – open, flat or closed – correspond, in general relativity, to three different geometries of space. Space is three-dimensional, and most people (including myself) cannot easily picture what these different geometries mean in three-dimensional space. But the salient features of the different geometries can be understood by looking at a two-dimensional surface.

A flat two-dimensional surface is just that – a flat plane, like the page you are reading. A closed surface is one which folds back on itself, and has no edges – the simplest example is the surface of a sphere, like the imaginary balloon covered in paint spots that helps us to understand how galaxies move apart as the Universe expands. The surface of the Earth is a closed surface, even though it is not a perfect sphere. It has a finite area, and no edges. An open surface is a little harder to visualise. The best example is a so-called "saddle" shape, or the shape of a mountain pass. This never folds back on itself, but instead folds outwards. It extends forever, out to infinity, unless it has a definite edge.

The three-dimensional Universe must fit one of the equivalent geometries. We know that it is reasonably close to being flat, but we don't know whether it is just open or just closed, or exactly flat. Inflation explains how a universe born with any curvature is forced towards flatness by its phase of exponential growth.

Suppose the universe starts out very closed, before inflation. The analogy, in two dimensions, would be a small sphere, one on which the horizon would be very close to any hypothetical observer on the surface (an intelligent ant, maybe). It happens that on such a curved surface the angles of a triangle do not add up to 180°, as they do on a flat surface, but always to more than this. How much more depends on the size of the triangle in comparison with the size of the sphere on which it is drawn. You might try to measure the effect on Earth by drawing a huge triangle in the Sahara Desert and measuring how much the sum of its angles is more than 180°. On the Moon, the same size of effect would show up for a smaller triangle. It would be very easy for an intelligent observer on a small sphere to measure the curvature of that world, by simple geometry. Now imagine doubling the radius of the sphere a hundred times. This vast inflation stretches the horizon off into the far distance as the sphere grows. Any small triangle drawn on the surface – a triangle the same size as the ones drawn on the original, small sphere – would have angles adding up very closely to 180°; deviations from flat geometry would only show up now if triangles could be drawn big enough to cover a significant fraction of the hugely inflated surface – triangles themselves doubled in size a hundred times compared with the first triangles. In other words, inflation flattens the surface.

The same is true in three dimensions. It doesn't matter whether the universe is open or closed to start with. Inflation drives it towards flatness, so powerfully that, in fact, any deviations from flatness that were there to start off with are reduced by the *square* of the exponential growth term.

Inflation theory cannot tell us whether the Universe is open or closed, but in addition to explaining some puzzling features of the Universe it makes one, and only one, new prediction – that the Universe we live in sits so close to the dividing line between being open and closed that no human measurement will ever be able to distinguish the difference from a perfectly flat universe. In other words, as far as any measurements we can make are concerned, if the inflationary idea is correct then omega must be *precisely* equal to one.

It looks as if the inflationary models are telling us something important about the Universe we live in, even though the precise details of the inflation mechanism are far from being worked out. But you ought to be puzzled about one aspect of this description of the Universe. If it was born in a chaotic, pre-inflation state, how did it come out of the big bang with enough order for there to be a well-defined arrow of time today, as entropy works to return everything to ultimate chaos? What wound the clock of the Universe up?

MAKING TIME

The evidence of the cosmic background radiation, and the uniformity with which it seems to fill space, is evidence that the Universe was in thermal equilibrium long ago. Today, we see bright stars in a dark sky. Even granted that there is an enormous amount of entropy in the background radiation, and that the stars represent only a minor deviation from equilibrium, in a thermodynamic sense, this is still a puzzle. To understand what is going on in thermodynamic terms, we first have to understand how that starlight is being made at all.

Stars shine because they are hot, and they are hot because, deep in their interiors nuclei of light elements, especially hydrogen, are being fused into nuclei of heavier elements, notably helium. In the process, some of the original mass of each of the fusing nuclei is turned into energy, following $E = mc^2$. A nucleus of helium-4, for example, has slightly less mass than two protons and two neutrons in isolation. The fusion process simply represents the natural tendency of all things to seek out a lower energy state, and in the process to increase the entropy content of the Universe. More complex fusion reactions inside stars build up modest amounts of still heavier elements, such as the carbon which is the basis of living things, and the oxygen in the air that we breathe. These heavy elements are scattered when dying stars explode, and some of this stellar debris will eventually coalesce with other interstellar material to form new planets circling new stars, and living creatures on those planets. But all that need concern us now is that when light nuclei fuse into heavier nuclei, energy is released.

Even though helium nuclei represent a lower energy state than hydrogen nuclei, the fusion process can only occur where the pressure and temperature are very high, because there is a barrier preventing two positively charged nuclei from "touching". The barrier is the electric repulsion force, which I have already mentioned. Something has to squeeze the two fusing nuclei together to the point where the short-range, but powerful, strong nuclear force can take over and overwhelm the electric force. It happened, for a brief interval, in the big bang itself. An individual star may be a less spectacular nuclear pressure-cooker than the big bang, but it has the merit of lasting for a very long time – billions of years of steady heat and pressure inside a star like the Sun. So, over a long period, the fusion process proceeds even further than in the big bang, making heavier elements than helium.

But if a stellar interior is like the later stages of the big bang, what were the earlier stages of the big bang like? Then (or there), conditions were even *more* extreme. At a certain temperature and pressure, simple nuclei that collide with one another will tend to stick together, because of the strong nuclear force. But what if the nuclei are moving even faster, so that the collisions are more violent? – which is what a higher temperature and pressure imply. Under those conditions, there is a tendency for any more complex nuclei to be smashed apart by the collisions, broken down into the basic units of neutrons and protons. At even greater temperatures and pressures, collisions between neutrons and protons are so violent that even these particles are smashed into their constituent quarks.

So the thermodynamic equilibrium of the early Universe was one in which not only were particles and radiation in equilibrium with each other, through the pair production process, but no complex nuclei could exist, in spite of the attraction of the strong nuclear force. In those conditions, there would have been no time asymmetry if it were not for the expansion of the Universe. The original arrow of time came simply from the expansion, pointing from the hotter past into the cooler future.

As the Universe expanded and cooled, it reached a state where stable nuclei could form, and the low energy state offered by nuclei such as helium became available to the protons and neutrons. But only a minor proportion of these particles were able to take advantage of the lower energy state before the expansion and cooling proceeded to the point where collisions between particles no longer surmounted the electric barrier. A non-equilibrium state was set up in the material world, directly as a result of the expansion of the Universe. The nuclear reactions going on inside stars provide the mechanism by which the Universe is trying to restore precise thermodynamic equilibrium in spite of the expansion.

There is still, though, another facet to the puzzle. How did stellar interiors get hot in the first place? Once there is enough heat for nuclear fusion to begin, the fusion reactions can keep things hot for as long as the fuel – hydrogen – lasts. Initially, though, a star starts out in life as a cloud of cool gas and dust in space. When such a cloud is pulled together by its own gravity, it becomes denser and more compact. The atoms at the heart of the cloud, squeezed by pressure from above, become hot; gravitational energy from the contracting mass of the cloud is turned into thermal energy, until the temperature rises to the point where nuclear fusion can begin.* Clearly, there is a profound link here between gravity and thermodynamics. That link was put on a secure footing in the 1970s, by the work of a mathematical physicist studying the equations that describe the most extreme manifestations of gravity at work – black holes.

---

* This is, of course, like the expansion out of the big bang in reverse. Expanding things cool down, and their energy is stored as gravitational potential energy; collapsing things hot up, as that gravitational potential energy is released and converted into heat.

THE BLACK HOLE CONNECTION

A black hole is a region of spacetime completely dominated by gravity. The escape velocity from such an object is so great that not even light can escape from a black hole, which is what gives them their name. The name is, however, slightly misleading, since if any black holes exist in our Universe they may well be surrounded by dense clouds of swirling gas and dust, attracted by the powerful gravitational field of the hole and falling into it. As always, matter falling down under the attraction of gravity and being squeezed into a smaller volume gets hot. The swirling cloud of material surrounding a black hole which contains as much matter as a star (a "stellar mass black hole") will get very hot indeed, radiating energy across the electromagnetic spectrum, from radio waves to X-rays and gamma rays. Indeed, astronomers believe that one or two black holes have been detected in our own Milky Way Galaxy, and perhaps in its nearest neighbours, by precisely this kind of energetic radiation. But the thermodynamics of black holes can best be understood in terms of more subtle effects at work.

There is a sharp "edge" to a black hole, called an event horizon. Anything inside the event horizon is trapped, and can never escape. Anything outside the horizon can, if it has enough speed, escape from the gravitational clutches of the hole, out into the Universe at large. The area of this surface around the black hole, the area of the event horizon, is the most important property of the hole, and represents its size; it is, of course, related to the mass of the hole, and also to whether it is rotating, and by how much. I shall discuss only the simplest case, a spherical, nonrotating state. The implications are fascinating, and mindstretching enough without going in for unnecessary complexities.

Stephen Hawking, of the University of Cambridge, realised that there is a profound connection between the description of a black hole in terms of general relativity and both thermodynamics and quantum theory. In so doing, he linked the two greatest developments of twentieth-century physics with the great achievement of nineteenth-century physics.

The key to this is the contribution from quantum physics, and in particular uncertainty, the principle that lies at the heart of quantum theory. There are several pairs of conjugate variables in the quantum world, and in every case it is impossible for both members of the pair to have a precisely defined value at the same time. Position and momentum are the archetypes – a quantum particle *does not have* both a position and a momentum. Similarly, the parameters energy and time are conjugate variables, and although in some ways the implications are even harder to grasp than the uncertainty of position and momentum, they are at least as important.

One way to interpret the energy/time uncertainty is to imagine a tiny volume of space, anywhere in the Universe. Ignoring any photons of the background radiation that happen to be passing through, we would say that this little bit of space contains no energy. But can we be sure? The uncertainty relation tells us that this little volume *might* contain a certain energy, E, provided that it only does so for less than a certain time, t. The relation between E and t is precisely defined by the quantum rules, in such a way that the bigger E is, the smaller t must be. A little bubble of energy can pop up, and promptly vanish, without ever being detected. Since energy can be converted into mass, this tells us that particle-antiparticle pairs can pop up, out of nothing at all, in the vacuum of empty space – provided they promptly annihilate one another.

Quantum physicists have a saying, that anything that is not expressly forbidden by the rules of quantum theory *must* happen. Fleeting pair production is not forbidden, so it is inevitable. According to quantum theory, the Universe is seething with these "virtual" particles, appearing and disappearing in their droves all the time. And their presence is even revealed, indirectly, by a minor influence on the way ordinary particles interact. That influence is a measurable, though indirect, confirmation of the existence of virtual particles that can never be directly observed,

because their lives are so short. It is one of many successful predictions that give physicists faith in the quantum
theory.

But if the whole Universe is filled with such particles,
and what we think of as the vacuum is a seething maelstrom of virtual pair production, what happens near a
black hole? Hawking's stroke of inspiration was to imagine
a virtual pair being produced in this way right on the edge
of a black hole, a minuscule distance above the event horizon. This must be happening all the time, all over the
surface, but we only consider one pair for simplicity. The
two particles in the pair will have their own momenta,
with velocities constrained by the rules of quantum physics
but which are briefly real. And it is possible for one of the
particles in the pair to be moving in to the horizon while
the other moves outwards. In less time than it takes for
the pair to annihilate, one has vanished forever into the
black hole, and the other is still at large in the Universe.
The uncertainty rules seem to have been broken, since a
particle has been created out of nothing at all, as far as
the outside Universe is concerned. In fact, Hawking showed
that the quantum rules are not violated, and that the
energy mc^2 of the "new" particle has come from the black
hole itself, which has in effect converted part of itself into
a real particle in the outside world. This process must be
going on all over the surface of an isolated black hole,
which must therefore leak away its mass, slowly, as a flood
of fundamental particles. Hawking proved that this cascade
of particles from its surface defines a temperature for any
black hole, a temperature which depends on its mass. Mass
is a relativistic property, linked with gravity; temperature
is a thermodynamic property; the two are related by
quantum physics.

The effect works in such a way that smaller black holes
are hotter. A hole containing a billion or so tonnes of
matter (less than one trillionth of the mass of the Earth)
would, for example, have a temperature of 10^{12} K, whereas
one with a mass ten times that of our Sun would have a
temperature of only 10^{-7} K.

As particles escape from a black hole, it gets smaller,
and hotter. So the "evaporation" proceeds ever more

rapidly, until eventually the hole gives up its last dregs of
mass in a burst of energy. But "eventually" can be a very
long time. A black hole which started out from the big
bang with a mass of 100 million tonnes, and has been
quietly evaporating ever since, without gathering any
more matter by gravitational attraction, will just about be
ready to explode now; a black hole with the mass of our
Sun will not evaporate completely until 10^{66} years have
passed – that is, 10^{56} times the present age of the Universe.

The implications of black-hole evaporation provide a
wealth of material for theorists to ponder. Observers have
even found some occasional strong bursts of radiation
coming from the depths of space, which just might be a
result of relatively low mass black holes giving up the
ghost, although nobody is betting on it. But all of these
ideas and speculations are of far less significance than the
discovery which lies at the heart of this work, Hawking's
discovery that there is indeed a fundamental relationship
between general relativity (that is, gravity), thermodynamics and quantum physics. This is not a relationship
which has been fully worked out and explored yet, by any
means. Even dealing only with general principles, however,
it gives us yet another insight into the thermodynamics of
the Universe and the nature of time.

Hawking's work tells us that gravitational fields have
entropy. One of the physicists who have tried to work out
the implications in the context of the expanding Universe
is Paul Davies, of the University of Newcastle upon Tyne,
who has related them to inflation and the arrow of time.
Any gravitational field has a well-defined entropy, which
is low when things are uniform and smooth, but high
when things are crumpled together tightly, that is, when
space, as described by Einstein's equations of general relativity, is highly crumpled up. When matter clumps into
galaxies and stars, and ultimately into black holes, it is
increasing the entropy of the Universe. Roger Penrose, of
Oxford University, stresses that the big bang was a very
special, smooth singularity, low in entropy. But the big
crunch, if it occurs, will be a very general, more messy
kind of singularity made from merging black holes, high

in entropy. On this picture, entropy *is* always increasing, even though the early Universe was in thermodynamic equilibrium. So, Davies says, we have another manifestation of the asymmetry of time, and the new question to be answered is: why did the Universe emerge from the big bang in a smooth state, gravitationally, whcn thermodynamics favours a highly chaotic, crumpled state, with some regions expanding, some collapsing, and black holes evaporating and exploding all over the place?

Perhaps, he says, that *is* the way things started out. In such a chaotic universe, there would inevitably be some regions, some tiny grains, that were just ripe for the kind of inflation that I have already described. The spacetime of our region of the universe has been "wound up" by inflation, in the same sense (but even more dramatically) that the asymmetry represented by nuclear fusion in stars was "wound up" in the later, more sedate expansion of the big bang itself. But the *direction* of the entropy flow doesn't depend on inflation. As Davies puts it:

> The remaining history of the universe is the subsequent attempt to unwind by gravitational clumping (galaxies $\longrightarrow$ stars $\longrightarrow$ black holes) and nucleosynthesis (hydrogen $\longrightarrow$ helium $\longrightarrow$ iron). Together these two evolutionary chains account for all the observed macroscopic time asymmetry in the world, and imprint upon our environment a distinct arrow of time.*

But this isn't quite the end of the story. Stephen Hawking has gone one step further. He has produced a theoretical model – a set of equations – that describe how a universe can come into being from nothing at all, *before* the inflationary era. His models include inflation as a necessity, and they also require that the Universe *must* be closed, with omega just a little bit bigger than one.

---

* *Nature*, volume 301, page 398, 1983; the ideas are developed further in *Nature*, volume 312, page 524, 1984.

REMOVING THE BOUNDARIES

For most of us, imagining the creation, if only for a fleeting instant, of an electron and a positron out of nothing at all is far from easy. But cosmologists, to paraphrase Lewis Carroll, have no trouble believing three impossible things before breakfast. In the early 1970s one of them, Ed Tryon, suggested that the entire observable Universe might be no more, and no less, than a single vacuum fluctuation of this kind.

His idea rests upon the way particles may be created out of singularities, states of infinite density like the singularity at the birth of the Universe, and can disappear into singularities, like those at the hearts of black holes. There is no evidence that our Universe contains equal proportions of matter and antimatter – indeed, one of the triumphs of modern cosmology is the way it explains how a remnant of one baryon for every billion photons came out of the early Universe. So the analogy with the production and annihilation of particle/antiparticle pairs is not exact. But Tryon pointed out that whereas the lifetime of such a virtual pair is limited by the amount of energy stored in their mass, it is possible for a universe to be created, out of nothing at all, with *no* net energy.

This is a neat trick. It rests upon the way energy is stored in a gravitational field, which is negative in the sense that the mass energy mc^2 of a particle is positive.* In effect, this means that there might be a different kind of vacuum fluctuation, in which there is a trade-off not between particles and antiparticles but between mass-energy and gravitational energy. Just as before, the lifetime of such a fluctuation depends on its overall energy, and the less energy it contains the longer it can live, according to the quantum rules. But what if it contains no energy at all? Why, then it can live forever!

---

* I go into the details of Tryon's work, and this unlikely-sounding suggestion, in *In Search of the Big Bang*.

Tryon suggested that our Universe might be such a bubble, created in a superdense state, out of nothing, with a mass-energy almost exactly balanced by the gravitational energy associated with that mass-energy. It can live for as long as you like, by fine-tuning the balance, even though it must one day collapse back into the nothing from whence it came. His idea didn't catch on at the time, because it contained one glaring flaw. Even though the quantum rules might allow such an enormously dense vacuum fluctuation to form, and would permit it to live forever, there is nothing in those rules which says it *must* live forever. And if the entire mass-energy of the Universe were concentrated in a tiny initial seed, then according to the standard ideas of the 1970s it ought to collapse, very promptly indeed, under its own gravitational attraction, and wink out of existence every bit as quickly as a virtual particle/antiparticle pair. But the idea has been rescued and refurbished in the 1980s thanks to inflation. With inflation, any tiny, superdense seed that comes into existence, no matter how it forms, can be whooshed up to macroscopic size before it has a chance to collapse; once it is the size of a grapefruit, the more modest expansion we see today will suffice to keep it growing for billions of years, even if it must ultimately collapse back into a fireball, then a seed, and then nothing at all. So Tryon's ideas, speculative though they are, have now been taken up by other researchers who are trying to find precise mathematical descriptions of how nothing can become something.

Hawking has come at the problem from a slightly different direction, and has come up with a model which is superficially not unlike this picture of the Universe. But there is a fundamental difference. Instead of worrying about how nothing became something at the moment of creation, he has tried to do away with the moment of creation altogether.

His reason is that a moment of creation provides an "edge" to the Universe, a boundary in time. If the Universe is closed, there is no boundary to space, any more than there is an edge to the surface of the Earth. The edge in time is called a singularity, and mathematicians abhor singularities – any theory that contains singularities is usually assumed to be flawed, so why should general relativity be an exception? Hawking has tried to do away with the singularity, improving general relativity by including aspects of quantum physics in the description of the Universe.

When cosmologists use the analogy of the Earth's surface as being "like" a closed universe, they say that the closed universe, of course, has one extra dimension. But Hawking goes further in his quantum treatment of the moment of creation:

> When quantum mechanics is taken into account, there is the possibility that the singularity may be smeared out and that space and time together may form a closed four-dimensional surface without a boundary or edge, like the surface of the Earth but with two extra dimensions. This would mean that the universe was completely self-contained and did not require boundary conditions ... there would not be any singularities at which the laws of physics would break down. One could say that the boundary condition of the universe is that it has no boundary.*

The importance of Hawking's model is that by removing the singularity it means that it is possible to describe the entire Universe in accordance with the known laws of physics. Physics can't cope with infinities, or singularities, but he has done away with both by removing the "moment of creation" with its implication of a state of infinite density in the beginning of the Universe. This is a great benefit to achieve simply by stretching our minds to encompass a closed surface with two dimensions more than the surface of a sphere, instead of one extra dimension; and it also has other implications. First, as by now you might expect of all good theories of the Universe, Hawking's version has an inbuilt phase of exponential growth which provides all the usual benefits of inflation.

* Quote from Hawking's essay "The Edge of Spacetime" in William Kaufmann's book *Universe*.

Secondly, the universe *must* be closed, in the three-dimensional sense that I have already used the term, or the trick with boundary conditions does not work. As far as we are concerned, living in three dimensions and observing the Universe around us with our senses and our scientific instruments, the Universe expands away from a superdense state, halts, and then collapses back into a superdense state. In Hawking's universe, omega is inevitably bigger than one. That universe also has a curious thermodynamic behaviour, which raises some of those unanswered questions that I warned you about.

THE REVERSING UNIVERSE

In Hawking's universe, time has no unique meaning at what we are used to thinking of as the moment of creation. He makes an analogy with the surface of a sphere, like the surface of the Earth, and points out that even though there is no edge to such a surface, the way we measure directions becomes confused at the poles. At the North Pole, there is no direction "north", nor indeed "east" or "west"; all directions away from the pole point to the south. At the moment of creation, there is no time "past", and all arrows of time point outwards into the future. The future is the direction in which the Universe is expanding. On Earth, we have a similar confusion at the South Pole, where all directions are now "north". But the analogy with time in Hawking's universe breaks down, because the thermodynamic arrow of time must *still* be pointing in the direction of expansion. Hawking can extend his analogy between the Universe and the surface of a sphere to describe expansion and collapse, but he can only do so if the arrow of time reverses at the moment of maximum expansion.

The analogy he uses is that this is like a journey from the North Pole to the South Pole. Everywhere down to the equator, the universe is expanding – successive lines of latitude get bigger. The arrow of time always points the same way. But from the equator onwards, successive lines of latitude drawn around the sphere get smaller, and the Universe shrinks. Nevertheless, says Hawking, time still points away from the pole and towards the equator – in the opposite sense to the way it pointed in the northern hemisphere. Or, to put it in slightly more "scientific" language, the Universe is finite in both space and time, and is time-symmetric, mirroring its own behaviour on either side of the moment of maximum expansion.

This is quite different from Penrose's picture of the progression from big bang to big crunch as a smooth thermodynamic process, with entropy increasing from a special initial state into one of a confusion of complicated black hole interactions. On Hawking's picture, the big crunch is *exactly* like the big bang in reverse, and is also very smooth. So entropy must *decrease* as the Universe shrinks. From our perspective, the contracting half of such a universe would be strange indeed. Instead of nuclear fusion reactions producing energy which makes stars hot and releases photons into space, there would be a flow of photons *out* from cold surfaces, across space and down on to the surfaces of the stars. The arriving photons would combine with one another in just the right way to break up complex nuclei into their constituent parts. On the surface of a planet like ours, the actions of wind and weather would conspire to build mountains out of sediment, with rivers running backwards. And the behaviour of living things would be even more bizarre. The processes we think of as decay would act in reverse, drawing together scattered material to form the living body of an old animal – such as a human being – who would grow younger as time passed, and whose bodily functions would be almost too bizarre to contemplate.*

It *sounds* crazy. But as Paul Davies has stressed, it is odd that the description should be so laughable, since it is simply a description of our present world in time-reversed

---

* Almost, but not quite. Philip Dick had a brave stab at it in his science fiction story *Counter-Clock World* (Berkley Medallion, New York, 1967).

language. Not only is the world of the contracting universe no more remarkable than our everyday world, it *is* our everyday world. The difference in the description is purely semantic. Furthermore, says Davies:

> A human being in a reversed-time world would also have a reversed brain, reversed senses and presumably a reversed mind. He would remember the future and predict the past, though his language would not convey the same meaning of these words as it does to us. In all respects his world would appear to him the same as ours does to us.*

In thermodynamic terms, any intelligent beings occupying either half of the universe will "see" the flow of time as from a more dense state to a less dense state, as Hawking requires. In each half of the universe, the inhabitants will think that they are living in the first half, the expansion phase, and that this will be followed by a collapse. In that sense, such a closed universe contains *two* beginnings, and no end!

The questions raised by this kind of model are uncomfortable. *How* does time reverse when the universe is at a state of maximum expansion? Does it happen suddenly, all over the universe, at the same instant? How can it know when to do so, everywhere at the same instant, if no signal can travel faster than light? Or could there be a transition period, in which time runs slower and slower, stops, and then reverses? What would that mean in thermodynamic terms? Either prospect worries many physicists, who find these problems sufficient reason to discard the closed universe idea and to accept the idea of a unique big bang, with an eternal expansion and a unique arrow of time pointing always in the same direction, not one which circles back on itself. But then they are left with the singularity, the edge of spacetime, and the puzzle of the moment of creation itself, and what went "before".

Which set of puzzles you are happier to live with is largely a matter of personal choice at present. My own preference is for Hawking's universe, with no edges. And one of my reasons for preferring the idea that the Universe

---

* *Space and Time in the Modern Universe*, page 196.

is just closed, rather than just open (remember, either way omega *must* be very close to one) is that there is yet another asymmetry in nature, an electromagnetic arrow of time, that also points in the direction of expansion.

ABSORBING THE PAST

Like Newton's laws of motion, the equations that describe the behaviour of electromagnetic radiation have no inbuilt arrow of time. They work as well describing events that move backwards in time, from our perspective, as they do describing events that flow forwards in time. The best way to see this is by looking at the electromagnetic radiation in terms of waves. Quantum theory tells us that at this level of reality it is possible to treat such phenomena either in terms of waves or as particles (photons), depending on circumstances; the wave description, which is appropriate here, uses a set of equations discovered in the nineteenth century by the Scot James Clerk Maxwell, and named, in his honour, Maxwell's equations. Among other things, they describe how the signal from your local TV station gets from the transmitter aerial to the antenna on the roof of your house.

These equations describe changing electric and magnetic fields, moving through space at the speed of light. To get a picture of what is going on, imagine a stone dropped into a still pond. Ripples spread out across the pond, away from the point where the stone was dropped in. In a roughly similar fashion, a wire which carries an alternating electric current, or a TV or radio mast driven by such a current, radiates electromagnetic waves outwards in all directions. It takes a certain amount of time for the wave to reach a point in space away from the wire, or antenna, and so physicists call this kind of behaviour "retarded" wave motion. Maxwell's equations describe, perfectly, the way waves of this kind propagate. But they do more.

There are, in fact, two sets of Maxwell's equations

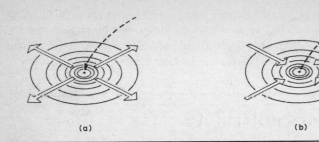

(a)   (b)

Figure 3.1/Time Asymmetry in Waves
We are all familiar with the waves that spread outwards
when a pebble is tossed into a pond (a). They are called
retarded waves. The laws of physics also allow for the
existence of advanced waves, which could converge on
a point and give up their energy to shoot a pebble high
in the air (b). The absence of advanced electromagnetic
waves in the Universe is another manifestation of time
asymmetry, and may be intimately connected with the
ultimate fate of the Universe.

(strictly speaking, two sets of *solutions* to those equations).
The second set describes the time reversed version of the
picture I have just painted, in which electromagnetic
waves move in towards a wire (or antenna) from the far
reaches of the Universe, converging perfectly in step and
combining to create an alternating electric current in the
wire. Because such waves disturb space far from the wire
before they reach the wire itself, they are called "advanced"
waves. We never see advanced waves in the real world,
but the symmetry of Maxwell's equations implies that both
are equally valid. The absence of advanced waves is a
feature of our Universe that defines an electromagnetic
arrow of time, seemingly independent of the thermodynamic arrow, but pointing in the same direction. And, it
turns out, the fact that we never see incoming radiation
of this kind is telling us something fundamental about the
nature of our Universe and its fate.

I have described the two solutions to Maxwell's equations in terms of waves moving out from, or coming in to,
a source. But another way of looking at this is to say that
the advanced wave is moving "outwards" from the source
but "backwards" in time. This is a better way of looking at

things, since it emphasises the fact that the electric current in the wire is indeed the source of the wave, its reason for existence. Many physicists are happy to dismiss the absence of advanced waves in our Universe simply on these grounds – "everybody knows" that effects always *follow* their causes, and, indeed, this is a principle dignified by the name "causality". Causes come first, effects later, and so "of course" we don't see advanced waves. But this is no more than semantics, raising the question *why* do effects always follow causes in our Universe? We are back, straightaway, to the arrow of time!

In the 1940s, two American physicists, John Wheeler and Richard Feynman, were investigating the problem of providing a good mathematical description of the way in which charged particles, like electrons, interact with electromagnetic fields. Their efforts met with only partial success, but led them to develop a mathematical treatment of advanced and retarded waves which gives full weight to both sets of solutions to Maxwell's equations, and lays the responsibility for the electromagnetic arrow of time squarely on the structure of the Universe at large. The Wheeler-Feynman theory has never made a major impact on science, although it is often discussed as an intriguing byway of mathematical physics. Perhaps, though, in the light of recent developments in cosmology it is time it was taken more seriously.

When an alternating electric current generates electromagnetic waves, it does so because electrons in the wire carrying the current are accelerated. In this case, they are being jiggled back and forth, but they would also generate an electromagnetic wave if they were accelerated continuously in a straight line, or round a large loop. Equally, a charged particle that "feels" the passage of an electromagnetic wave will move in response. For simplicity, the Wheeler-Feynman approach can best be understood in terms of the behaviour of a single charged particle (perhaps an electron) being accelerated in the Universe, and the response of all the other charged particles in the Universe to the wave generated by that electron.

Since there is no asymmetry in Maxwell's equations, Wheeler and Feynman took both the advanced and

retarded solutions and combined them, in equal proportions, to provide a description of the way the accelerated
charged particle interacts with the electromagnetic field.
The accelerated electron creates, in equal measure, a
retarded wave moving outwards into the future, and an
advanced wave moving back into the past. This advanced
wave will arrive at other charged particles, and make them
move, *before* the original electron is accelerated. But, and
now things begin to get complicated, this means that the
other particles are accelerated, all at different times, depending on when the waves reach them, both before the
original electron is accelerated and, a suitable time later,
when the retarded wave eventually reaches them, *after* the
original electron is accelerated.

Each charged particle in turn, when it is given a jostle
by either the advanced or the retarded wave, will itself
generate both advanced and retarded waves. The result,
from the acceleration of one electron for a short time, is a
complex sea of overlapping electromagnetic waves, both
retarded and advanced, spreading out from all the other
charged particles and moving both forwards and backwards in time.

But all these waves have a common origin, in the
motion of the original electron, and are therefore very
similar to one another. They will interfere with one another, following well-established mathematical rules, in
the same way that the ripples from two pebbles dropped
into a pond will interfere with one another to produce a
new pattern of ripples. And all of this complexity of
interactions will happen instantaneously, from the point
of view of the original electron, as a simple example
makes clear.

Suppose another charged particle is just far enough
away from the electron that the electromagnetic wave,
travelling at the speed of light, takes one hour to cross the
gap. This will trigger a response from the second particle,
part of which will be in the form of an advanced wave,
moving backwards in time at the speed of light, and
arriving at the original electron one hour before it left the
second particle – just at the instant that the original
electron is radiating! The same logic applies to all particles,

at all distances from the original electron. How, in these
strange circumstances, will the two waves (and all the
other waves from all the other charged particles) interact?
The great achievement of the work by Wheeler and Feynman was to show, using some rather hairy mathematics,
that all of the advanced waves cancel out, under suitable
circumstances. The advanced waves from the rest of the
charged particles in the Universe not only cancel out the
advanced waves from the original electron, they exactly
double up the strength of its retarded wave, so that instead
of a 50:50 contribution from each of the solutions to
Maxwell's equations we are left with a single, full-size contribution from the retarded wave alone. But the cost of
doing this places restrictions on the Universe which made
the theory look unattractive, until recently.

The Wheeler-Feynman solution to the puzzle of why we
don't see advanced waves works perfectly if the universe
they are describing with their equations is mathematically
equivalent to a closed box, like the closed systems beloved
of thermodynamicists. The opaque walls of the box provide
the exact response required to cancel out the advanced
waves and boost the retarded waves to full strength; but
the trick doesn't work if the box is open and energy can
get out of it. If radiation can escape from an accelerated
particle and disappear into space without ever meeting
another charged particle, then, clearly, there is no scope
for the production of the advanced waves from the future
of that particle which will cancel out its own advanced
waves. In technical terms, for the Wheeler-Feynman trick
to work there must be a "perfect absorber" in the future of
our Universe – the Universe itself must be a "closed box".

What happens to the radiation moving outwards from
all the stars in the Universe today in such a closed universe? During the contracting phase of the universe it
must, of course, converge on to cold stars, heating them up
and driving nuclear reactions backwards, and so on, exactly
in line with the description of the strange behaviour of a
contracting universe with time flowing backwards. This is
no coincidence; the Wheeler-Feynman description of electromagnetic radiation works only in a universe where, as
in Hawking's universe, the arrow of time reverses when

the state of maximum expansion is reached. And the same
reversal of the arrow of time explains why it is the
advanced wave that cancels out and the retarded wave
that survives in our Universe. What matters is that *one* of
the two solutions to Maxwell's equations cancels out; we
see the surviving wave as the retarded wave in an expanding Universe. From our point of view, the surviving
wave could be the advanced wave in a collapsing universe.
But, just as before, any intelligent beings in *either* half of
the Universe will see both the thermodynamic and electromagnetic arrows pointing in the direction of expansion,
with retarded waves being normal.

During most of the 1960s and 1970s, however, cosmologists favoured the idea that our Universe is open and
will expand forever. They did so largely because the
amount of matter we can see in stars and galaxies is not
enough to close the Universe. So the Wheeler-Feynman
"absorber" theory was dismissed as flawed, in some as yet
undiscovered way. An example of the way people thought,
quite recently, can be found in Paul Davies's book *Space
and Time in the Modern Universe* (pages 186 and 187):

> The ever expanding Friedmann models [of the Uni-
> verse] are inconsistent with this opaqueness require-
> ment. The recontracting model is, however, perfectly
> opaque to radiation. The current evidence in favour
> of a low density, ever-expanding universe should
> therefore be considered as evidence against the
> absorber theory.

This appeared in print as recently as 1977. But since
then the inflationary models have appeared, with their
resolution of the deep puzzles of the uniformity and flatness
of the Universe, and Hawking's work has provided a sound
theoretical basis for expecting the Universe to be closed,
and therefore opaque. From the perspective of the mid-
1980s, ten years on from that comment by Davies, the
success of the Wheeler-Feynman theory in explaining the
electromagnetic arrow of time in a closed universe is just
one more piece of evidence to add to the growing weight
which tells us that our Universe *must* be closed, with
enough mass in it to ensure that omega exceeds one, if

only by the tiniest amount. But still, as Davies correctly
pointed out ten years ago, there is a wealth of evidence
that all the baryons in all the stars in all the galaxies do
not provide enough mass to close the Universe. Where is
the missing mass? What is it? And how much do we need?
To answer the last question first, we need to find out just
how much, or how little, baryonic material – matter in
the form of protons, neutrons and everyday atoms – there
is in the Universe.

CHAPTER FOUR

ELEMENTARY
EVIDENCE

Astronomers have known for half a century that there is more to the Universe than meets the eye. In the early 1930s, the Dutch astronomer Jan Oort was one of the pioneers who deduced the nature of the Milky Way Galaxy by studying the way the visible stars are moving. It was only at about that time that these measurements showed conclusively that the stars of the Milky Way are each in orbit around a centre quite distant from the Sun, moving in a way reminiscent of the way planets orbit around the Sun. Our Solar System lies about two-thirds of the way out from the centre of this swirling system, in the galactic suburbs. In our neighbourhood, motions of nearby stars can be studied in some detail. They do not move perfectly in a single plane, but wobble up and down as they orbit the centre of the Galaxy, moving a little above and a little below the main plane of the Galaxy. The height to which a star moving at a certain velocity can climb out of the plane before it is pulled back down by the gravity of all the other material in the plane depends, of course, on the overall mass of the disk in the neighbourhood of that star. The more mass there is in the plane, the more tightly each individual star is held down to the plane by gravity. And by studying the distribution of the stars around the plane of the Milky Way, Oort showed that there must be three times as much matter in the solar neighbourhood as we can see in the form of bright stars.

Of course, he couldn't watch a single star moving up and down through the plane. These changes take thousands or millions of years. But the overall distribution of stars, the relative numbers at each distance above and below the centre of the plane, can be determined and compared with distributions deduced from the laws of orbital dynamics. These numbers give a reliable picture of the way gravity is constraining the movement of the stars. This kind of study shows that the stars are being held in place by several times more material than we can see in the bright stars themselves. Since the 1930s, about as much mass as there is locked up in the visible stars near our Sun has been identified as cold clouds of gas and dust spread between those stars, but that still makes up, together with the stars themselves, only two-thirds of the amount required to explain the local dynamics of the Galaxy.

MASS AND LIGHT

This unseen, dark matter can be measured in terms of a number called the mass-to-light ratio, or M/L. This is defined to be 1 for our Sun – one solar mass of matter, in the form of a star, produces one solar luminosity of light. In the region of the Galaxy near the Solar System, Oort's figures tell us that M/L is about 3. That doesn't seem a very dramatic discovery, but at about the same time that Oort was finding evidence of dark matter ("missing mass") close to home in the Universe, Fritz Zwicky, a Swiss astronomer who made a lifelong study of distant galaxies, found evidence of dark matter on a much more impressive scale.

Zwicky was studying clusters of galaxies, groups containing several systems like our own Milky Way Galaxy which lie together in space. Our Galaxy is a member of a small cluster called the Local Group; it has only a handful of members. Some clusters contain hundreds of galaxies. Astronomers assume that these clusters are groups in which the galaxies are kept together by gravity, orbiting around each other but moving through space as a group, rather like a swarm of bees. But when Zwicky used the ubiquitous Doppler shift to measure the velocities of individual galaxies in one group, the Coma cluster, he found that they were moving much too rapidly, relative to one another, to be held together by the gravitational pull of all the stars in all the galaxies of the cluster. It looked as if the flying galaxies ought to have moved apart, dissolving the cluster, long ago when the Universe was young. And he found the same thing when he looked at other clusters – they are all full of galaxies moving much too fast to be held together by the gravity of the stars we can see.

There are many uncertainties in this approach. The masses of the galaxies, for example, can only be estimated from their brightnesses, on the assumption that an average star in a distant galaxy is as bright as an average star in our Galaxy. The distances to the clusters themselves are uncertain, which also affects the argument. But the size of the effect that Zwicky found, and has since been confirmed by every similar study, is so big that it dwarfs any possible errors of this kind in the calculation. In round terms, the amount of matter needed to stop clusters of galaxies from evaporating away is so great that M/L rises to about 300 – there is *three hundred* times more dark matter in clusters of galaxies than there is in the form of bright stars. For comparison, the amount of matter required to close the Universe, distributed uniformly through space, is only three times this. If the Universe is closed, its overall M/L is about 1,000.

None of this worried astronomers very much in the 1930s, or even in the '40s, '50s and '60s. The nature of the expanding Universe, and even the fact that the Universe extended far beyond our own Milky Way Galaxy, were new ideas to astronomy fifty years ago, and there was ample room for speculation about how these observations might be resolved. On the small scale of Oort's discovery, it hardly seemed reasonable to believe that astronomers had found every kind of object that might exist in the Milky Way, and it was easy for astronomers to imagine that there might be many very faint stars ("brown dwarfs") or objects like large planets ("Jupiters") contributing plenty of mass, but very little light, to the Milky Way. These assumptions have, as we shall see, been borne out by recent discoveries. Zwicky's evidence was more puzzling, but in the absence of any evidence to the contrary the theorists could speculate that the space between the galaxies within clusters might be filled with a sea of gas, enough to hold the galaxies together by gravity. These assumptions have not been borne out by later observations, but the pioneers weren't to know that. It was only when the big bang theory became established as a good description of the real Universe that the puzzle of the missing mass came to the forefront of astronomy. For, by a nice stroke of irony, it turned out that what had seemed in the 1940s to be a failure of the big bang model was, in fact, providing such a profound insight into the nature of the Universe that since the 1960s it has been able to tell astronomers, very precisely, how much matter there "ought" to be in all the stars and galaxies – at least, in the form of baryons.

MAKING ATOMS

Our everyday world is made of atoms. These come in many varieties, or elements – hydrogen and oxygen (sometimes combined to make up molecules of water), carbon, iron and the rest. Both the physical world and life itself depend on the interplay of atoms, combining in different ways and interacting to produce substances as diverse as the DNA which carries the genetic code in our cells and the gold that we value for its pretty colour and scarcity. But where do the atoms come from, and why is gold scarce on Earth, while water is abundant? These seemingly philosophical questions can be answered in great detail by astronomers, using the big bang model of the Universe.

Our home planet is far from being a typical part of the Universe. In terms of the atoms from which it is made, it is far from being a typical part even of our own Solar System. By far the bulk of the matter in the Solar System is concentrated in the Sun, around which all the planets orbit. The Sun alone contains as much matter as 333,400 planets like the Earth, and all the planets of the Solar System put together contain less than 450 Earth masses – less than 0.15 per cent of the mass of the Sun. The Sun is a much more typical representative of the Universe than the Earth is, and our Sun seems to be basically similar to the thousands of millions of other stars that make up our Milky Way Galaxy, which is itself basically similar to the hundreds of millions of galaxies that make up the visible part of the Universe. The Sun and stars do *not* contain the same elements, in the same abundances, that we find on Earth.

Astronomers can find out what stars are made of by looking at their light. Each type of atom – each element – produces its own characteristic pattern of lines in the spectra from the stars, and the relative strength of these lines shows the proportion of each element present. Using radio astronomy techniques, it is even possible to probe the composition of cool clouds of gas between the stars in this way, and by and large the picture that astronomers get is always the same, wherever they look in the Universe. The great majority of all the material in all the stars and clouds of all the galaxies is in the form of hydrogen, the simplest element of all. A significant amount of the material in stars (about 25 per cent) is in the form of helium, the next simplest atom, but only a few per cent of the material in any star is in the form of heavier elements like the carbon, oxygen, iron and the rest that are so important on Earth. In some stars, less than one-hundredth of 1 per cent of all the material present is in the form of heavy elements; all the rest is hydrogen or helium. And the stars which contain least in the way of heavy elements are always the ones that seem, from other evidence, to be the oldest stars in the Galaxy.

All this must be telling us something profound about the nature of the Universe. Hydrogen, after all, is the simplest element – an atom of hydrogen consists of a single proton associated with a single electron. In the most common form of helium, two protons and two neutrons together make up the nucleus of the atom, with two electrons outside it. Most of the Universe is made of the simplest kinds of atoms. Inside a star, the electrons are stripped away from the nuclei, and lead a more independent existence, but it is still nuclei of hydrogen (protons) and of helium (also known as alpha particles) that make up most of the mass of the visible universe.*

In the 1940s, George Gamow, a Russian-born physicist who spent most of his working life in the United States, suggested that *all* of the heavier elements in the Universe had been built up from hydrogen nuclei – protons – in the big bang itself. At that time, physicists were just coming to grips with the idea of nuclear fusion, in which two light nuclei fuse together to make a heavier nucleus, and release energy in the process. Fusion of hydrogen into helium is the energy source both of a star like the Sun and of the hydrogen bomb, but this is simply the first step up a fusion "ladder" that could in principle reach up to carbon, oxygen and other heavy elements. In order to do the trick, you need to pack the light nuclei together in a very hot, dense ball – and where better, thought Gamow, than in the fireball of the big bang?

Unfortunately, when he and his colleagues carried the calculations through, they found they couldn't make the heavy elements in the big bang after all. The fireball simply expanded too quickly. Starting out with hydrogen (protons), there was a burst of fusion in the fireball that turned 20 to 30 per cent of the matter into helium, agreeing rather nicely with studies of the composition of stars. But before the later stages of the fusion process could get going to make the kind of atoms our bodies are made of, the universe had expanded and cooled to the point where no more fusion could occur.

---

* Because they combine to form atomic nuclei, protons and neutrons are together also known as nucleons. They are the baryons from which atomic nuclei are made. There are other kinds of baryons, which can be manufactured in colliding beam experiments at particle accelerators, but they do not occur in significant numbers in the Universe at large today. For the purposes of the present discussion, the terms "nucleon" and "baryon" are interchangeable.

The failure was not an embarrassment for long, since in the 1950s Fred Hoyle and his colleagues showed how the heavier elements could indeed be produced by nuclear fusion, in just about the right proportions, by reactions going on inside stars. Stars don't get as hot as the big bang, and nuclei are not packed so tightly together inside stars as they were in the big bang, but stars have the great virtue, for a nuclear pressure cooker, of lasting for a long time – millions, even thousands of millions, of years. The slow processes of nuclear fusion *can* turn hydrogen into helium, helium into carbon, carbon into oxygen, and so on up the ladder, inside stars. When some stars reach the end of their lives, they explode, sometimes creating even heavier elements in the process, and scatter their products into space where they form the clouds from which later generations of stars are made. On this picture, it is easy to see why the oldest stars (made from original material left over from the big bang) contain only hydrogen, helium and a tiny smattering of heavier elements, while younger stars, like our Sun, made from the debris of earlier stars, contain a richer mixture of heavy elements and can have a retinue of planets, amounting to no more than a fraction of 1 per cent of the mass of the star itself. Only on some of those planets, where most of the hydrogen and helium has blown away into space, do the heavy elements seem, to local life-forms inhabiting those planets, to be the most important constituents of the Universe.

All the atoms that are important to life on Earth, except hydrogen, were made inside stars that blazed in the Milky Way before the Sun was born (helium is *not* important to life on Earth). We know where atoms come from, and how they are made. But Gamow's idea that nuclei might be manufactured in the big bang (big bang nucleosynthesis) bore unexpected fruit in the 1960s, when cosmologists at last realised just how good a description of the Universe the big bang model is.

BARYONS AND THE UNIVERSE

When gas expands, it cools. When a hot fireball of radiation expands, it, too, cools. From one point of view, you can think of this in terms of photons acting as the "particles" of the radiation gas, losing energy like the atoms of an expanding gas cloud. For radiation, however, the effect of this loss of energy is to change the wavelength of the radiation, redshifting it by an appropriate amount. Energy that was in the form of gamma or X-rays, or even more energetic forms, is progressively redshifted down through the ultraviolet and the visible spectrum, on into the infrared and then into the radio wavebands. The exact temperature of this kind of radiation can be related to the exact way its energy is distributed at different wavelengths, and to the value of the wavelength for which the radiation has a peak intensity. It is called blackbody radiation, and the way in which its energy is shared among the wavelengths is called the Planck distribution, in honour of Max Planck, one of the founders of the quantum theory.

In the 1940s and 1950s, cosmologists who toyed with the big bang idea, playing intellectual games with the concept of a definite beginning to the Universe, realised that the hot fireball of the big bang should have left its trace today in the form of a cold, blackbody background radiation with a temperature of a few degrees above the absolute zero of temperature, a few K. But no serious effort was made to search for this radiation, which probably tells us just how much (or how little) faith astronomers really had in the big bang theory at that time. When the background radiation was discovered in the 1960s, as a weak hiss of radio noise at microwave frequencies, coming from all directions in space, it was an accidental discovery by radio astronomers actually working on a different problem. But the penny soon dropped, and over the past twenty years many observations of this background radiation at many wavelengths have shown how precisely it follows the blackbody form of the Planck distribution, for radiation with a temperature of 2.7 K. It is *possible* to imagine ways – rather contrived ways – in which the Universe could have been filled with radiation like this without a big bang origin, but the simplest explanation of the background radiation is that there was indeed a big bang. It was this discovery which gave the impetus to cosmology that has led to the modern version of the big bang model, which makes very precise "predictions" about just how much of the lightest elements should have been synthesised in the big bang itself.

A few seconds after the moment of creation, with the temperature of the Universe about 10 billion K, the fireball was already laced with a trace of baryonic material, in the form of neutrons and protons. With electrons also present, the two kinds of nucleon were to some extent interchangeable, since a proton and electron can be forced to combine, at high energies, into a neutron, while a neutron left to its own devices will decay into a proton and an electron. The first process, combining protons and electrons to make neutrons, became less and less common as the Universe cooled, and by the time the temperature was down to about a billion K, when the Universe was roughly 3 minutes old, the balance had shifted to the point where there were 14 protons and just 2 neutrons in every 16 nucleons. At this temperature, an individual proton and an individual neutron can stick together to form a nucleus of deuterium, or heavy hydrogen. When the temperature was higher, deuterium nuclei could form but were immediately smashed apart by high energy photons; at "only" a billion K, however, the deuterium nuclei not only hold together themselves but promptly combine in pairs to make nuclei of helium-4, each containing two protons and two neutrons. Out of every 16 nucleons, 4 have gone into helium – 25 per cent. The rest are left as protons, to become hydrogen atoms as the Universe cools further and they capture electrons. The presence of some 25 per cent helium in the oldest stars is the strongest evidence, together with the background radiation itself, that the big bang model is a good description of how the Universe we live in came into being.

In fact, though, the nuclear fusion reactions are a little more complicated than I have indicated. Instead of two deuterium nuclei combining directly to make one nucleus of helium-4, it is more likely that they will interact to produce a helium-3 nucleus and a lone neutron. The "spare" neutron can interact with another helium-3 nucleus, almost immediately carrying the reaction through to helium-4. And that is still not quite the end of the story, since although there are no stable nuclei which contain either five or six nucleons, the fusing together of nuclei of helium-3 and helium-4 can produce a little lithium-7 before the Universe has cooled to the point where no more fusion reactions occur. Within four minutes of the moment of creation, however, all of this activity is over, and the primordial element abundances have been decided.

If we could measure the abundances of all of these elements in the oldest stars, then we would know the mixture that came out of the big bang. Because the exact proportion of each element produced depends on, among other things, the density of the fireball in which the elements were made, that could, if the Universe were made only of baryons, immediately tell us whether the Universe is open or closed, and therefore its ultimate fate.

Of all the lightest elements, the production of helium-4 is least sensitive to the density. But the amount of helium-4 produced does depend critically on the rate at which the Universe was (and is) expanding, so that measurements of helium-4 abundances are still of vital importance. It is easily the most accurately measured of all the abundances of elements heavier than hydrogen, and those measurements just fit with the requirements of the standard model of the big bang. The best modern estimates give a proportion of about 23 to 25 per cent of all the material in old stars as helium-4. Deuterium is equally interesting in principle, but very difficult to measure in practice. Deuterium is not manufactured inside stars at all, according to the modern understanding of nuclear physics. At the temperatures inside stars deuterium is actually destroyed. So any measurement of the abundance of deuterium in the Universe today must give a figure lower than the actual abundance which came out of the big bang. The measurements are difficult, but such techniques as measuring the amount of deuterium in samples from meteorites, and spectroscopic studies of the clouds of Jupiter, indicate that for every hundred thousand atoms of hydrogen in the Universe there are only two or three atoms of deuterium. Astronomers believe that perhaps twice as much deuterium, five atoms for every hundred thousand atoms of hydrogen, came out of the big bang, and that the rest has been destroyed inside stars.

Spectroscopic studies of the light from stars suggest that helium-3 was produced in roughly equal amounts to deuterium in the big bang, and lithium-7 is even rarer, with perhaps five atoms of lithium-7 having been produced in the big bang for every 10 billion atoms of hydrogen. All of these values for the element abundances can be explained in terms of the standard model of the big bang – provided that the total density of baryons in the Universe is significantly less than one-tenth of the critical value needed to make the Universe closed.

In its simplest form, the argument can best be seen in terms of deuterium. Deuterium nuclei in the early Universe are very likely to collide with one another and produce helium nuclei. When the density of nuclei is low, when they are few and far between, there will be relatively few collisions and so proportionately more deuterium will survive to be detected today. When the density of nuclei is high, there will be relatively more collisions, and fewer deuterium nuclei left for us to see. Even five deuterium atoms for every hundred thousand hydrogen atoms represents a high figure, on this basis. The present measurements of deuterium abundance set a very strong limit on the density of baryons in the early Universe, and therefore in the Universe today, and this limit is far below the critical value.

When the calculations of big bang nucleosynthesis were first carried through in this form in the 1960s, they seemed to provide conclusive evidence that the Universe was open, and would expand forever. That remained the view of cosmologists into the 1970s, for it never occurred to anyone to suggest seriously that the Universe might be largely made up of other forms of matter than the ones we see around us. It was taken for granted that baryons were the most important form of matter in the Universe. Indeed, not just baryons, but visible baryons, in the form of bright stars and galaxies. As recently as 1981, when I discussed the problem with John Huchra, of the Smithsonian Institution Observatory, he commented. "From a philosophical point of view, an optical observer would quickly lose interest in observational cosmology if the Universe was dominated by things he couldn't see." Well, Huchra's philosophical objections to the existence of large amounts of dark matter in the Universe seem to have been overruled – and the observers don't seem to have given up their trade, but are busy studying the dynamics of the visible galaxies in order to probe, indirectly, the distribution of the dark matter in the Universe. And most of that dark matter probably is not baryonic. The supposition that the Universe is dominated by visible stars and galaxies now seems to have been based on nothing more than a kind of baryon-chauvinism. Just as the fact that elements such as carbon, oxygen and iron are common on Earth does not mean that they dominate the composition of stars and galaxies at large, so the fact that stars are composed chiefly out of baryons does not necessarily mean that the Universe at large is composed chiefly out of baryons. It *might* be; but already the figures for mass to light ratios are suggesting otherwise.

Translating the constraints on baryon density set by the measured abundances of the lightest elements into these terms, M/L *must* be less than 72 for the whole Universe if most of the mass of the Universe is in the form of baryons.* But we already know that for clusters of galaxies M/L is much bigger than this, 300 or so. The natural conclusion is that there is a great deal of mass "out there" that is not in the form of baryons – and that is the conclusion that cosmologists have been reluctantly forced to in the past few years. Having overcome their initial reluctance, however, they now seem ready to embrace the idea wholeheartedly. After all, if there must be *some* non-baryonic matter, and there may well be a great deal of it, there might even be enough to push the value of omega close to

* The figure comes from the work of David Schramm, of the University of Chicago, who presented it to a meeting of the Royal Society, in London, in October 1982.

one, where the best modern cosmological theories say it ought to be.

In order to find out what this dark matter might be, and how it has affected the evolution of galaxies and clusters of galaxies themselves, we have to step into the strange world of the particle physicists. Fortunately, however, the first step into that realm does not take us too far away from the familiar world of baryons.

BEYOND THE BARYONS

We now know that the elements are not really elementary. Atoms are made of baryons (specifically, protons and neutrons) and electrons. Because the mass of an electron is very much smaller than that of a nucleon, ordinary matter is often referred to as baryonic matter, and in calculating the density of the Universe on the basis of the way the lightest elements were synthesised in the big bang I didn't bother to add in the mass of all the electrons. This is reasonable enough, since although there is one electron in the Universe for each proton, the mass of each proton is roughly two thousand times the mass of each electron. In everyday units, the mass of an electron is 9×10^{-28} of a gram – that is, a decimal point followed by 27 zeros and a 9. Everyday units are a bit impractical for measuring such small masses, and physicists prefer to use a unit called the electron Volt, or eV. Strictly speaking, this is a unit of energy. But mass and energy are interchangeable, since $E = mc^2$, and the c^2 term is usually taken as read. In these units, the mass of an electron is a little over half a million eV, written as 0.5 MeV, and the mass of a proton is 938.3 MeV, while the mass of a neutron is 939.6 MeV.

These three tiny particles are the building blocks of everyday matter, all of the atoms in your body and in planet Earth, the material that makes up the Sun and all the bright stars we can see in the sky. But there is one more component of everyday matter that we have not yet met. It is called the neutrino.

In the 1930s, physicists realised that there must be a fourth kind of particle involved in nuclear reactions. When a neutron decays into a proton and an electron, for example, the energy of motion carried by the particles, as well as their mass energy, can be measured by suitable experiments. Whenever this was done, the experimenters found that the total energy being carried by the proton and the electron (mass energy + energy of motion) was less than the total energy carried by the original neutron. Something had to be carrying energy away. That "something" was called a neutrino, by Wolfgang Pauli, as long ago as 1931. But neutrinos were not directly detected by any experiments until 1956.

The reason it took so long to detect neutrinos directly is that they interact only very weakly with ordinary matter. Neutrinos that are produced in nuclear reactions in the heart of the Sun, for example, pass through the whole thickness of the Sun itself more easily than light passes through a sheet of glass. Neutrinos are important in nuclear reactions, and anywhere that the density of matter or energy is very great, like the centre of a star or the big bang itself. But they pass through ordinary matter such as planets and people as if it wasn't there at all. But still, neutrinos are necessary to balance the books in nuclear reactions, and their existence has indeed been confirmed by experiments. They also complete a pleasing symmetry in the pattern of particles at the level important for nuclear physics.

Protons and neutrons are both members of the family called baryons, and it seems to be a fundamental law of nature that the number of baryons in the Universe today stays the same. When a neutron decays, it doesn't just disappear in a puff of energy, or turn into a different kind of particle; it turns into a different baryon. But along the way it seems to create an electron, so obviously electrons are not conserved in the way that baryons are. The presence of the neutrino, however, restores the balance. The electron and the neutrino are together members of a family, called leptons. And just as the number of baryons in the Universe is conserved, so is the number of leptons.

You can picture the decay of a neutron in two ways. In the first, a neutrino (lepton) is absorbed by a neutron (baryon) and this then decays into a proton (baryon) and an electron (lepton). One lepton and one baryon are converted into a different lepton and a different baryon. From the other point of view, a neutron on its own can decay to produce a proton (conserving baryon number) and also an electron plus an *anti*-neutrino. Just as electrons have antiparticle counterparts called positrons, so neutrinos have antiparticle counterparts. In adding up the total number of leptons in the Universe, you have to balance the books by subtracting the antiparticles from the particles. So by "creating" a lepton (electron) and an anti-lepton (anti-neutrino) the neutron decay has maintained the balance of leptons in the Universe. This is an important realisation, because nature does seem to favour this kind of symmetry. As physicists have probed deeper into the particle world over the past fifty years, developing mathematical rules which describe particle interactions ever more completely, they have found symmetry and balance at every level – for there are indeed levels, or at least one level, beyond the baryon.

All the evidence we have is that leptons are truly fundamental particles. There is no structure within an electron or a neutrino, and neither of them can ever be divided into smaller components. But baryons are different. In the 1960s, physicists established that the behaviour of baryons (and of other kinds of particle, which needn't bother us here) can best be explained if each proton and each neutron is made up of three more basic particles, called quarks. Just two types of quark are needed to account for the properties of nucleons. They have been given the arbitrary names "up" and "down", and a proton is best thought of as a tightly knit group made up of two up quarks and one down quark, while a neutron is made of two down quarks and one up quark. Other combinations of these two types of quark, including pairs of quarks, can, together with the lepton pair, explain all of the behaviour of everyday matter. Although individual quarks cannot exist in isolation, the up/down pair has many similarities in its basic properties to the electron/neutrino pair. Like

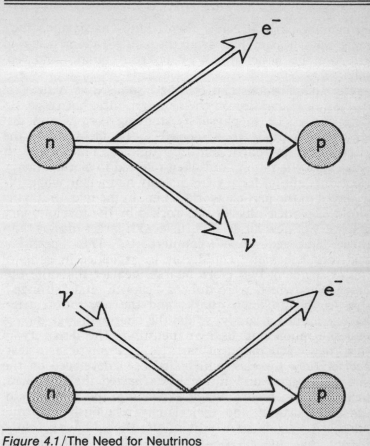

Figure 4.1/The Need for Neutrinos
When a neutron "decays" into a proton and an electron,
the equations can only be balanced by including another
particle, the neutrino. Strictly speaking, a neutron
decay produces a proton, an electron and an electron
anti-neutrino; alternatively, an incoming neutrino
colliding with the neutron is converted into an electron,
while the neutron is converted into a proton.

the leptons, quarks have no internal structure and seem
to be truly fundamental particles, and in the early 1970s
it seemed that physics was on the brink of explaining
the material world in terms of just four types of truly

elementary particle. Since then, things have got a little more complicated, but even the complications seem to preserve the basic symmetry between quarks and leptons.

The complications can best be visualised as nature repeating her quark-lepton theme – not once, but twice. As long ago as 1936, physicists studying cosmic rays discovered a particle that seemed to be identical to the electron except that it had 200 times as much mass. It was called the muon, and often referred to as the "heavy electron", but for forty years nobody had a clue what role it played in the particle world. Then, in the mid-1970s the world of particle physics was rocked by the revolutionary discovery of new kinds of particle with larger masses than their "first generation" counterparts.* These particles, and their behaviour, could only be explained in terms of the existence of two more quarks, heavier than the up/down pair, which were dubbed "charm" and "strange". Just as the up/down quarks and the electron/neutrino leptons formed a family, so did the charm/strange quarks and the muon and its own neutrino. The discovery of this "new" generation of particles gave physicists a test-bed for their theories, which had been developed on the first generation, and the theories passed the tests with flying colours, predicting, or explaining, the properties of the new particles and their family relationships to one another.† Nobody knew why nature should have repeated herself, but the physicists were delighted at the opportunity to confirm and refine their models. Scarcely had they done so, however, when they had more to think about.

If one heavy electron can exist, why not more? With

---

* These particles are actually *manufactured* out of energy in machines that accelerate everyday particles, such as protons and electrons, to very high energies and smash them into one another. The more massive families of particles would exist naturally in a high energy world (or early in the big bang), but they do not occur naturally in the Universe at large today, and when they are made artificially they soon decay into more familiar particles of the everyday world.

† Evidence for the "strange" quark had actually come in the 1960s. But it was only in the 1970s that physicists appreciated the need for "charm" as well, and realised that nature was duplicating the basic quark/lepton theme.

particle physics in turmoil in the mid-1970s, a team at Stanford University searched for evidence of a superheavy electron, and found it: a "new" lepton, identical to the electron except for its mass, an impressive 2,000 MeV – an "electron" that weighs twice as much as a proton! The new particle was given the name tau, and, as expected, turned out to be partnered by its own type of neutrino. There were now *three* generations of lepton, but only two generations of quark. The implications were obvious, and the search for a third generation of yet heavier quarks was on. Traces of one, called bottom, were found in 1977; evidence for the existence of the other, called top, has come from experiments at CERN in the mid-1980s. The quark-lepton symmetry is restored, those favoured theories of the physicists have been tested yet again and still come up trumps, and once again we have a balanced picture in which now six kinds of quark and six kinds of lepton are needed to account for everything in the material world, while only two of those quarks and two leptons are needed to account for all the matter in the everyday world. But where will it all end? If nature can indulge herself by allowing heavier copies of the two basic quarks and the two basic leptons, and can then make yet a third generation of particles, given enough energy, perhaps the process is endless. Why stop at three generations, or four, or five? Just when it seems we have got too deep into the particle swamp to have any hope of escape, however, cosmology comes to the rescue. There *are* good reasons why we should not expect the particle physicists to discover, or manufacture, any more generations of heavier quarks and leptons – and those reasons have to do with the influence of neutrinos on the way the lightest elements were synthesised in the big bang.

NEUTRINO COSMOLOGY

The "extra" generations of particles do not exist in the Universe today, except where they are made in high energy events. When they are made, they soon decay into the familiar particles of the everyday world. So there is no possibility that particles made of the heavier quarks (charm, strange, top and bottom) can provide a significant amount of extra mass in the Universe, over and above the mass of the baryons produced out of the big bang.* But the extra generations of particles can exist where there is enough energy, and there was certainly enough energy for them to have played a part in the processes going on in the big bang, just before the time, about three minutes after the moment of creation, when the baryons settled down into a mix of 75 per cent hydrogen, 25 per cent helium-4 and a trace of deuterium, helium-3 and lithium-7. The existence of the heavier quarks does not affect the subsequent story of the Universe at all, and for practical purposes we can ignore them. But the presence of three kinds of neutrino during the early stages of the big bang does turn out to be very significant in determining how fast the Universe expands, and therefore influences its ultimate fate.

The important point is the number of distinct varieties of light particle that exist. "Light", in this context, means with a mass of less than about a thousand electron Volts, 1 keV. The electron neutrinos were originally thought to have no mass at all, and certainly fit this requirement; more recently, as we shall see, there have been suggestions that the electron neutrino may have a mass of a few eV, and that there may be enough of them to provide all the

---

* Perhaps I should say "very little" chance. Some physicists have speculated that a form of matter containing equal numbers of up, down and strange quarks might be stable. If so, and *if* such matter was produced in the big bang before the epoch of baryon formation, then there *might* be enough of this "strange matter" about today, perhaps in the form of "quark nuggets", to play a part in providing the dark matter. But I wouldn't bet on it.

missing mass. This no longer looks the best bet in terms of dark matter candidates, but neutrinos are, indeed, very light. Such light particles are also sometimes referred to as "relativistic" particles, because they emerge from the big bang with very large velocities, close to the speed of light,* which we know from relativity theory is the ultimate speed limit in the Universe. By analogy with the rapid motion of energetic particles in a hot gas, they are also referred to as "hot" particles. In that terminology, neutrinos left over from the big bang which carry a trace of mass would represent the presence of hot, dark matter in the Universe.

But that is getting ahead of the story. For primordial nucleosynthesis, especially of helium-4, what matters is that the three kinds of neutrino are indeed distinct. This shows up in experiments here on Earth. When I said earlier that a neutrino can interact with a neutron to produce a proton and an electron, I should have specified that an *electron* neutrino was involved in the interaction. If a muon neutrino interacts with a neutron, then, as we might expect, the interaction produces a proton plus a muon. *How* the three flavours of neutrino "know" which partner they belong to physics cannot, yet, tell us. But they do know, and the flavours are different. Armed with that information, cosmologists can set very precise limits on the amount of helium-4 that could have been produced in the big bang.

The amount of helium-4 produced by nucleosynthesis in the big bang does not depend critically on the density of the universe. Roughly 25 per cent helium emerges from the big bang for a wide range of possible densities. But the amount of helium-4 produced does depend on how quickly the universe is expanding at the time of nucleosynthesis, in such a way that *more* helium-4 is produced if the universe is expanding faster. This is because if the universe is expanding rapidly, more neutrons can be locked up in helium nuclei before they have time to decay into protons. If the universe expands slowly, more neutrons decay before the temperature falls to the point where helium nuclei are

---

* *Exactly* the speed of light, if their mass is precisely zero.

stable. The rate at which the universe is expanding at the
time of nucleosynthesis depends on how many different
kinds of relativistic particle there are in the universe at
that time. You can think of this as like a pressure forcing
the universe to expand – gas under pressure in a cylinder
will force a piston to move outwards as the gas expands,
and the more gas there is packed into the cylinder to start
with the more pressure there will be and the faster the
piston will move. The analogy isn't exact – in the early
universe, this particular "pressure" does not involve the

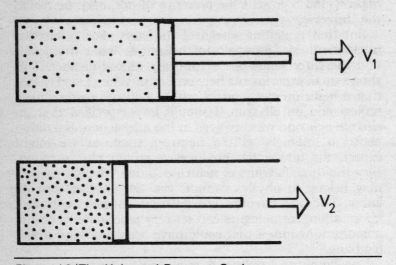

Figure 4.2/The Universal Pressure Cooker
The more tightly compressed a piston is, the harder the
trapped gases push outwards. Let the piston go, and the
pressure will make it move faster (v_2) than if the gas is
less compressed (v_1). The speed with which the
Universe expanded during the big bang depends on how
many neutrinos (among other things) were doing the
"pushing"; the amount of helium manufactured in the
big bang depends on how long the "pressure cooker"
régime lasted. By balancing the two requirements and
measuring the proportion of helium in old stars today,
astrophysicists deduce that there are certainly no
more than four families of neutrinos, and that the
three kinds already discovered are probably all there
are to find.

massive particles, such as protons, but only the light particles of the relativistic "gas". These include photons, electrons and positrons, and the three types of neutrino and their antineutrino counterparts. And the predictions for helium-4 abundances that come out of the calculations are astonishingly precise.

We know for sure, of course, that the photons and electron/positron pairs were present in the big bang. So, taking those as read, we can express the predicted helium–4 abundances in terms of the number of neutrino flavours present. With only two types of neutrino (and their antineutrino counterparts) allowed for in the calculations, less than 23 per cent of the baryonic matter can be in the form of helium-4. With three types of neutrino, the proportion of helium-4 rises to 24 per cent, while with four neutrino flavours the predicted abundance is rather more than 25 per cent. In round terms, the addition of each extra flavour of neutrino increases the amount of helium compared with hydrogen by one percentage point. The latest measurements of helium abundance in the real Universe give a figure of 23 to 25 per cent, exactly in agreement with the prediction if there are three neutrino flavours only. The figures can just be stretched to include the range corresponding to four flavours, but no further; the best evidence from cosmology is that all the varieties of neutrino that exist in the Universe have already been identified.

The extraordinary power of this cosmological insight into the world of particle physics can be seen by looking at how difficult it has been for the particle physicists to decide, on the basis of their experiments carried out here and now on Earth, how many (or how few) neutrino flavours there ought to be. In the early 1980s, using various indirect arguments, and a lot of wishful thinking, the physicists could only say that there must be less than 737 varieties of neutrino. Over the next few years, they struggled to push this limit down, first to 44, then to 30 flavours. In the past couple of years, there have been hints (only hints) from high energy experiments at CERN that the limit could be set as low as six or seven flavours. But cosmology got there first, and even in 1986, at the time of writing, the best limit on the number of neutrino flavours comes not from experiments here and now on Earth, but from astrophysical measurements of helium abundances in distant stars and our understanding of nucleosynthesis long ago in the big bang. The beautiful way in which everything hangs together is persuasive evidence for both cosmologists and the particle physicists that the standard model of the big bang is correct, that there are three flavours of neutrino in the Universe, and that the amount of baryonic matter in the Universe is only enough for an omega of about 0.1.

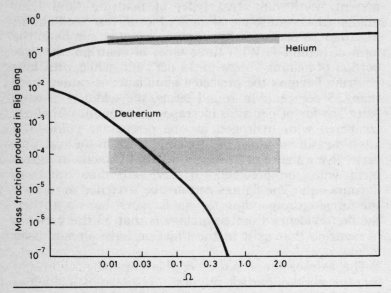

*Figure 4.3/*The Missing Mass
Standard models of the big bang set limits on how much helium and deuterium can have been produced, as a fraction of the amount of hydrogen, assuming most of the mass of the Universe is in the form of baryons. The shaded boxes show the range of options that is in agreement with observations of the Universe today. The evidence tells us that there is no more baryonic matter in the Universe than about one-tenth (0.1) of the amount required to make the Universe closed, with $\Omega = 1$. Yet other evidence shows that the Universe is very flat, and contains much more matter than this. The missing mass cannot be baryons; what is it?

The bulk of the dark matter required to close the Universe (in the favoured cosmological models) or to hold large clusters of galaxies together (based on dynamic evidence) is not baryonic. So what is it?

COLD, DARK MATTER

The limit set by cosmology and the helium abundance on the number of neutrino flavours is actually even more stringent than I have so far spelled out. The fact that the actual helium abundance in the Universe is about 25 per cent (this is the "primordial" abundance, after allowance has been made for helium manufactured inside stars since the big bang) is consistent with their being only five kinds of relativistic particle at all playing any significant part in the expansion of the cosmic fireball at the time of nucleosynthesis. These are the photons, electrons/positrons and the three flavours of neutrino. The particle physicists tell us that there may be other kinds of particle that play a part in particle interactions at high energies today. Although they have not yet detected any of these particles directly, some of them seem to be needed in order to round out the symmetry of the equations in the most successful particle physics theories, and the theorists have happily given these hypothetical particles names, such as photino and gravitino. Cosmologists refer to them collectively, perhaps slightly tongue-in-cheek, as "inos".

Although the theories that predict, or allow, the existence of inos are regarded as the best theories of particle physics we have, they cannot as yet tell us exactly what masses all of the "extra" particles ought to have. Some estimates set the mass of the photino, for example, as low as 250 to 500 eV. That is low enough to make them relativistic, and involve them in the helium production process in the same way that the different flavours of neutrino are involved. If such particles do exist, then they must certainly have participated in the interactions going on in the fireball. But what cosmology, and the observed helium abundance, reveal is that there can only have been a limited number of such particles (not a limited number of flavours, now, but only a few actual particles) around at the time of nucleosynthesis, compared with the numbers of neutrinos, or they would have influenced the expansion rate of the Universe and led to the production of more helium than we can actually see today. The other relativistic (light) particles are still allowed to exist, but their abundances compared with neutrinos must be very low at the time if nucleosynthesis – "suppressed", as some cosmologists eloquently put it. And that means that even if these particles still exist in the Universe today, there may not be enough of them to contribute enough mass to push omega right up to one, and close the Universe.

As it happens, one way to "suppress" the number of such particles is for them to decay into something else, early in the big bang. The rate at which such particles decay depends, among other things, on their mass, and in order to make the photinos (or other inos) decay quickly enough they have to be assigned a mass of a few thousand MeV (a few GeV; that is, a few times the proton mass). In that case, of course, they wouldn't contribute in quite the same way to the expansion of the Universe. Particle physicists are quite happy to assign a mass of a few GeV to the photino. Quite apart from this cosmological requirement, some of the particle theorists do favour a mass close to that of the proton, to get the most symmetry in their equations, and the cosmology may well be telling them (and us) which of the options left open by particle theory is the right one to choose. But we don't have to worry too much about the particle physics here.* The important point is that the only relativistic light particles which are allowed to fill the Universe are the three flavours of neutrino. Photinos, and other inos, must either be present in very small quantities compared with neutrinos, or much more massive, or both.

The particle physicists, however, still have some tricks up their sleeve. The same supersymmetry theories also predict the existence of heavier particles, inos with masses from a few GeV up to a few thousand times the mass of

---

* Some of the details of the supersymmetry theories are given in the Appendix.

the proton. Like protons themselves, such heavy particles would not affect the argument about the expansion rate of the Universe and the production of helium, because they are cold – they do not move at relativistic speeds. And there is one oddball particle, called the axion, which particle theorists introduced in the mid-1970s to explain some of the features of the interactions they observe, rather in the way that the neutrino was first introduced as a hypothetical particle, to explain observations of the way neutrons decay. The axion could be very light indeed, perhaps with a mass of only one hundred-thousandth of an electron Volt. But the way in which it is created keeps it "cold", in the relativistic sense, so that axions do not have the high, relativistic velocities typical of electrons and the light inos (perhaps including photinos) in the early Universe, and do not influence the expansion rate.

There is no shortage of candidates for the dark matter now known to dominate the Universe. The nature of these particles, and the reasons why theorists think they must exist, are discussed in the Appendix. But until some experiment here on Earth does detect one of these dark matter candidates, we need not worry too much about exactly which ones are which. What matters is that they come in two principal varieties. First, there are the hot, dark matter particles. The constraint imposed by the helium abundance tells us that the *only* relativistic particles that can be important on a cosmological scale now are the three known flavours of neutrino – although other light inos might be important on smaller scales, perhaps in galaxies or small groups of galaxies. The electrons have long since settled down to a quiet life inside stars and clouds of gas and of dust, where their mass is included in our loose use of the term "baryonic matter", while the photons, although still relativistic, have no significant mass to contribute. If one or more of the neutrino flavours had even a small mass, that would certainly be sufficient to make the Universe closed, because there may be as many neutrinos of each flavour as there are photons in the Universe.

Then there are the cold, dark matter candidates, which include the one-off cold, light particle, the axion, and a

variety of more massive particles. Happily for the cosmologists, even without knowing exactly which particles
are involved, they can distinguish between the effects of
hot and cold dark matter in the Universe at large, by
looking at the dynamics of galaxies and clusters of galaxies.
Once again, as with the limit on the number of possible
neutrino flavours, it turns out that cosmology can tell the
particle physicists what their experiments here on Earth
are likely to find. But the dynamics of galaxies can only be
properly understood within the context of the dynamic
evolution of the whole Universe, how fast it is expanding
today and how old it is. So the search for the missing mass
now takes us back into the realms of "traditional" cosmology.

CHAPTER FIVE

DYNAMIC FACTORS

The only reason that astronomers are able to determine the properties of the Universe at large is that, relatively speaking, galaxies are much closer to one another than stars are. One of the best ways to picture this is by using an imaginary model of the Universe based on aspirins. If our Sun were the size of an aspirin, then the nearest star would be represented by another aspirin 140 kilometres away. This is fairly typical of the spaces between stars – the distance from one typical star to its nearest neighbour is several tens of millions of times the diameter of the star itself (except, of course, for binaries and similar systems where two or more stars orbit closely around one another). Galaxies, like our own Milky Way, contain thousands of millions of stars, spread over appropriately large volumes of space, but all held together, orbiting the galactic centre, by gravity. We can get an idea of the spacing between galaxies by changing our scale so that now the Milky Way, not the Sun, is represented by an aspirin. On this new scale, the nearest galaxy, M31, is represented by another aspirin, just 13 centimetres away.

This is slightly misleading, because both the Milky Way Galaxy and M31 are members of a small group of galaxies, called the Local Group, held together by gravity. The distance to the nearest similar small group of galaxies, the Sculptor Group, is still only 60 centimetres on the aspirin scale, however; and only three metres away, on this picture, we find the Virgo Cluster, a huge collection of about 200 galaxies, spread over the volume of a basketball. The Virgo Cluster is at the centre of a loose swarm of galaxy clusters that it dominates gravitationally; these include both the Local Group and the Sculptor Group, and the whole swarm is known as the local supercluster.

We can go on, on this picture. Just 20 metres away there is another big cluster, the Coma Cluster, containing thousands of galaxies. Further out, there are even larger clusters, some 20 metres across. The powerful radio-emitting galaxy Cygnus A is 45 metres distant; the brightest quasar on the night sky, 3C273, 130 metres away. And the entire visible Universe can be contained within a sphere roughly one kilometre across, on the scale where an aspirin represents our Galaxy.

It doesn't make much difference which of these distances you choose as representing a typical spacing between galaxies. Even the distance to the Virgo Cluster is only 600 times the diameter of our Galaxy; M31 is just about 25 Milky Way diameters distant from us. If galaxies were as far apart, relatively speaking, as the stars within galaxies, then the distance to our nearest galactic neighbour would be a hundred times further than the most distant object ever seen in the real Universe. Clearly, extragalactic space is far richer in galaxies than galactic space is in stars. And that enables cosmologists to get a broad picture of the way visible matter is distributed through the Universe, and how that distribution has changed as the Universe has evolved.

The most important observation in cosmology is that the light from all galaxies outside the Local Group is redshifted, which implies that the Universe is expanding and has evolved from a much denser state, the big bang. Hubble showed that for as far as it is possible to estimate distances to galaxies by other means (studies of variable stars or bright clusters of stars in galaxies, and so on) the redshift is proportional to distance. There seems no reason to doubt

that this rule applies to all distant galaxies, so distances to
galaxies far beyond the range of the variable star technique, or any other trick, can be determined simply by
measuring the redshift in their light and multiplying by a
constant, known today as Hubble's constant, H. Indeed,
even if we do not know the exact value of H, as long as
we know the rule "distance equals constant × redshift" we
can determine the *relative* distances to galaxies – that one
is twice as far away as another, while a third is, say, 12
times further away than the first. And that is just as well,
because estimates of the exact value of H have changed
significantly since Hubble's day, and even now there are
two schools of thought regarding the value that ought to
be assigned to Hubble's constant. Because all our estimates
of extragalactic distances are linked to Hubble's constant,
the effect is that one school holds that the observed Universe is twice as big as the estimate favoured by the other
school. Yet both groups base their estimates on the same
data, and each rejects the other claim as impossible. These
conflicting claims first emerged at a major scientific
meeting in Paris in 1976, and are still unresolved. But as
we shall see, only one of them is fully compatible with the
possibility that our Universe is closed, with omega close to
one.

STEPPING STONES TO
THE UNIVERSE

In comparison with the precision with which physicists
today know such fundamental numbers as the mass of the
proton, or the size of the constant of gravity, it may seem
remarkable that there could be as much uncertainty as a
factor of two in our knowledge of the distance scale of the
Universe. But this apparent vagueness becomes a little
more understandable when you realise that it was only in
the 1920s that astronomers appreciated for the first time
that there is more to the Universe than our own Milky

Way Galaxy, and began to estimate distances to other
galaxies. Indeed, compared with the best estimates available in 1929, the Universe today is ten times "bigger" than
it was – at least, it's ten times bigger than astronomers
thought it was.

The main reason for the imprecision of their estimates is
simply that you can't put the Universe in your laboratory
to study it. A proton can actually be manipulated in the
lab and its properties measured; but our knowledge of the
Universe depends on observations of faint and distant
objects, and is always at best secondhand. The wonder is
that any plausible numbers for such properties as the distances to galaxies and quasars emerge at all, and the
ultimate parameter, the distance scale that gives us the
size of the Universe, is reached only by using a series of
scientific stepping-stones, each of which can only be
reached with the aid of earlier steps. A mistake anywhere
in the chain of reasoning throws off all the calculations
down the subsequent steps.

The shorthand expression "size of the Universe" is something of a misnomer. What astronomers are interested in
is the bit they can see, with the aid of telescopes and other
instruments, and what they want to know is a way to
calculate the distance to every galaxy and other object they
see out beyond our own Milky Way. They prefer to talk
about the distance *scale* of the Universe, the relative distances between galaxies, precisely because these relative
distances stay the same whatever the actual value of H.

Hubble's constant is the key number in all of cosmology.
Armed with an accurate value of H and redshift measurements, it would be possible to calculate the distance to
any galaxy. And it is the precise value of H that has been
bitterly disputed by the experts for ten years. Allan Sandage, of the Mount Wilson and Las Campanas Observatory,
and his colleague Gustav Tamman of the University of
Basel, estimate it as 50 kilometres per second per Megaparsec (km/sec/Mpc). Gerard de Vaucouleurs of the University of Texas advocates 100 km/sec/Mpc. Neither
seems willing to budge. But even within that range of
possibilities H is telling us a great deal about the Universe we live in.

The time that has elapsed since the big bang depends on how fast the Universe is expanding – on Hubble's constant. So measuring Hubble's constant also gives us, immediately, an estimate of the age of the Universe. That estimate is always too big, because gravity must have slowed down the expansion as the Universe has aged, making H smaller today than it was in the past (which is why it is sometimes denoted by H_0, denoting the value of Hubble's "constant" today, and why some pedants prefer the term "Hubble parameter", since it is not really a constant). The rate at which the universal expansion is slowing down depends, of course, on how much matter there is in it. The more matter there is, the more strongly gravity is acting to halt the expansion. If the density of the Universe is just the minimum required for closure, and omega has a value of one, then the true age of the Universe, the time that has elapsed since the big bang, is exactly two-thirds of $1/H$.

Even if we do not know the precise value of H at least the limited range of possibilities is telling us something about the time that has elapsed since the big bang. The inverse of H is called the Hubble time, and, once all the kilometres and Megaparsecs have been divided into one another, and the seconds converted into years, this ranges from 10 billion years (for $H = 100$ km/sec/Mpc) to 20 billion years (for $H = 50$ km/sec/Mpc). The corresponding range of possible ages for the Universe if omega is equal to one is, in round terms, 6.5 billion years to 13 billion years. And the uncertainty arises because of the difficulty for astronomers in finding the accurate distance to just one galaxy outside the Local Group.

Astronomers are only able to measure distances directly for objects that are very close to our own Solar System, compared with the distance scale of our own Galaxy, let alone that of the Universe. The method used is called parallax, and depends on the apparent shift in the position of a nearby star against the background of distant stars as the Earth moves in its orbit around the Sun. This is a direct, trigonometric method that requires just two observations of the star made six months apart, and an accurate knowledge of the diameter of the Earth's orbit

around the Sun. The parameters of the Earth's orbit are
known quite well, from applying parallax within the Solar
System and using Kepler's and Newton's laws; I won't go
into the details here. But, just to be on the safe side, and
out of habit, astronomers prefer to give distances determined by the parallax technique – and, ultimately, *all*
astronomical distances depend on this technique – in terms
of the angle through which the target star seems to shift
as the Earth moves around the Sun. This gives the distance
in parsecs, for "parallax seconds of arc", which can be converted into linear distances, assuming that we do indeed
know the size of the Earth's orbit – 1 pc is 3.26 light years.
The method works out to a distance of about 30 pc, beyond
which the angular shift is too small to measure accurately.

The next step out into the Universe involves making
observations of whole clusters of stars, over periods of
many years. Such clusters contain hundreds or thousands
of stars, bound together gravitationally and moving as a
unit through space. Provided the cluster is close enough to
us for its movement to be detectable, the way the size of
the cluster changes, year by year, can give a handle on
the distance to it. The technique is, in essence, the same
as the one you use automatically when crossing the road,
to decide how quickly an approaching vehicle is moving,
and whether you have time to cross safely. Unfortunately,
this technique is less precise than direct parallax measurements. It involves measuring the radial velocity of the
stars in the cluster (the velocity along the line of sight),
using red or blue shifts, measuring the component of
velocity across the line of sight by watching the cluster
move, and relating these velocity measurements to the
changing appearance of the cluster to deduce a distance.
Only a couple of clusters are close enough to study in this
way – the rest are too remote for the necessary observations to be carried out. But one of these clusters, the
Hyades, provides the next step out into the Galaxy.

The moving cluster method gives the distance of the
Hyades as 46 pc, with an error estimated as plus or minus
10 per cent. The cluster is also just close enough for the
distances to some of its stars to be determined, approxi-

mately, by parallax, and such measurements confirm the distance. The next step is to analyse the overall properties of the cluster.

Stars shine because they are hot, and they are hot because nuclear fusion reactions – nuclear burning – occur inside them. For most of its life, a star like our Sun is burning hydrogen, converting it into helium, and this is a long-lasting process that gives such stars stability. The brightness of a star, its surface temperature and the colour of the light it radiates, depend only on its mass, as long as its supply of hydrogen fuel lasts. In a cluster of stars, like the Hyades, there will be many stars with different masses, but all roughly the same age, busily burning hydrogen in this way. Because both the brightness of a star and its colour (more specifically, the nature of its spectrum) depend on mass, when astronomers plot the observed brightnesses of stars in a cluster like the Hyades against their colour, or spectral type, they do not get a random scatter of points on the diagram, but a slightly curved band within which most of the points lie. This band is called the main sequence, and it covers the range of possible brightnesses and colours for all normal, hydrogen-burning stars.

Such a diagram – called a Hertzprung-Russell, or HR, diagram – can be plotted for any cluster of stars. The colours of the stars are not affected by distance, but, of course, their apparent brightnesses are. So the main sequence band on the HR diagram will fall lower down for more distant clusters. And because astronomers know the distance to the Hyades cluster, they can determine the distances to other clusters by calculating how far those clusters would have to be moved towards us in order to brighten them up just enough for their main sequences to lie exactly in the same part of the diagram as the Hyades main sequence.

We are now well into the realms of statistics, averaging over hundreds of stars, in effect, every time the technique is applied. The main sequence provided by those hundreds of stars in each cluster is regarded as one single light with a known intensity – a standard candle. And the main sequence method takes astronomers out to distances of several hundred parsecs from home, sufficiently far to calibrate the most important cosmological standard candle of them all.

Some clusters of stars – by no means all – contain one or more members of a family called the Cepheid variables. A Cepheid is both interesting and useful because each member of the family pulsates with a regular rhythm, and the period of the cycle of any particular Cepheid is related to its brightness (both, ultimately, depend on the mass of the star). This rule was discovered by Henrietta Leavitt, at

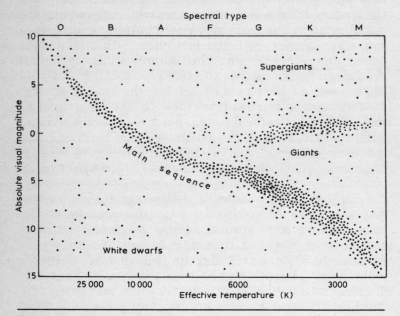

*Figure 5.1/*The H–R Diagram
The appearance of a star can be described in terms of its brightness (magnitude) and its temperature or colour (spectral type). A Hertzsprung-Russell diagram is a plot, like a graph, in which the position of each star is determined by these two properties. Most stars, burning their nuclear fuel in accordance with the simple laws of physics, lie on a band called the main sequence. Large, hot stars are to the top left of the main sequence; small, dim stars lie to the bottom right.

the Harvard College Observatory, in the early years of this
century. The discovery came from a study of Cepheids in
the Magellanic Clouds, which we now know to be satellite
galaxies of our own Milky Way. The stars in the Magellanic
Clouds are so far away that for most purposes they can
all be regarded as the same distance from us – the distance from one side of the cloud to the other is a small
fraction of the distance from us to the cloud. So the
period/luminosity relation showed up clearly in Leavitt's
data. It meant that if the distance to just one Cepheid
could be determined, then by using the period-luminosity
relationship and the apparent brightness of any other
Cepheid, the distance to that star could also be determined.

That is just what astronomers did, with the aid of the
main sequence method and the Hyades cluster. And, since
Cepheids are bright enough to be distinguished in nearby
galaxies, such as the Andromeda galaxy, they provided the
first means to measure the distances to those galaxies,
helping to establish that there are indeed other galaxies
beyond the Milky Way, and giving Hubble the basic information for his first estimate of the parameter we now
call H. That first estimate gave a rather embarrassing age
for the Universe, just under two billion years. Since geo-
physicists in the middle of the twentieth century already
had evidence that the age of the Earth is more than four
billion years, it was clear that something was wrong, but
the discrepancies were not resolved until 1952, when
Walter Baade, at the Mount Palomar Observatory, found
that there are two types of Cepheid, one much brighter
than the other. As a result, the accepted value of H was
reduced, and the calculated age of the Universe increased.

But even Cepheids only take us out to the nearby
galaxies, about 5 Mpc. Stellar explosions, novae and
supernovae, can be used as distance indicators to more
remote galaxies. A supernova shines, briefly, as brightly
as a whole galaxy of ordinary stars, and the apparent
brightness of such a flare tells us how far away the galaxy
in which the exploding star sits is – provided, of course,
that we have some idea of the absolute brightness of a
typical supernova explosion. Even then, supernovae are far

from common. So astronomers have to fall back on secondary distance indicators for any except the closest galaxies.

Secondary techniques are much less reliable. First, astronomers study the properties of those galaxies for which distances are known, and try to find common features. Then, by comparing those features with the equivalents in more remote galaxies, they estimate the distances to those galaxies.

For example, many spiral galaxies contain large clouds of ionised hydrogen, called HII regions. If all HII regions are the same size, and if it is possible to measure the diameters of these clouds using radio astronomy techniques, then the distance to a galaxy that contains such clouds can be found by comparing their apparent sizes with those of clouds in a nearby galaxy. There are plenty of "ifs" in that chain of reasoning, and other secondary methods are no better. Which is why it has proved possible for Sandage and de Vaucouleurs to hold such differing views on the exact value of Hubble's constant, and why I have gone into such detail in explaining the chain of argument used to build up an estimate of H.

There are several reasons why the two schools of thought disagree. First, de Vaucouleurs assumes that when we look out through the polar regions of our own Galaxy at distant galaxies then the light we see from them is dimmed slightly due to obscuring dust. Sandage disagrees, so his estimates of galaxy brightnesses, and distances, differ from de Vaucouleurs'. Secondly, Sandage has moved away from the simple period/luminosity relation for Cepheids found by Leavitt, and recognises slightly different period/ luminosity relations for Cepheids of different colours – an effect which de Vaucouleurs ignores. Among other differences, Sandage uses an estimate of 40 pc for the distance to the Hyades, much lower than the figure used by anyone else – and it is the distance to the Hyades which, through the main sequence method, gives the distance to the first Cepheid used to calibrate all the others. The two experts already disagree by more than 30 per cent in their distance estimates in our own astronomical backyard, and the discrepancies get worse as they move out beyond the nearer galaxies. Part of the further discrepancy in the estimates for distances out across the Universe lies in different interpretations of how much the cosmological motion of our Local Group is distorted by the pull of the Virgo cluster. Redshifts, and distances, to other galaxies have to be corrected for this local effect before we can get a true picture of how rapidly the Universe at large is expanding, and the two groups disagree on the size of the correction required.

There are new techniques being developed which may resolve the controversy. When a supernova explodes, it blasts out a shell of material, expanding very rapidly. The light from the supernova actually comes from this expanding shell, and the Doppler shift in that light tells astronomers how fast the shell is moving, which makes it a straightforward matter to calculate how big the shell is a certain time after the initial outburst. If there were some way to measure the apparent size of such a shell then this could be related to calculations of its actual size to give a direct, and theoretically well-founded, primary indication of distance.

The idea is simple, but the practical application is tricky. At the distance to the Virgo Cluster, for example, we are talking about measuring an angular size smaller than one millionth of a degree. Nevertheless, this astonishing accuracy is now being achieved by radio astronomers using the technique of Very Long Baseline Interferometry (VLBI). The first successful application of the technique in this way was announced in 1985, and gave the distance to a supernova in a galaxy called M100, 19 million parsecs away. The observations of the growing supernova shell suggest that the value of H is about 65 km/sec/Mpc, which probably would not please either Sandage or de Vaucouleurs, if taken at face value, but the uncertainties involved in this first application of a new technique are very large, and certainly encompass the range of values that would please either side in the debate. The method could, however, become one of the most reliable ways of estimating distances to galaxies, and thereby the distance scale of the Universe, in the next decade.

How else could the numbers be checked? The most promising line of attack for the immediate future is simply to carry out traditional observations of Cepheids in more remote galaxies. The Hubble space telescope, scheduled to fly on a shuttle mission in 1986, would have provided enough resolution to pick out individual Cepheids in the Virgo Cluster. Now, that will have to wait at least another year, but when the telescope does reach orbit it will also improve the accuracy of astronomical observations of all the other measuring techniques as well. Sandage and de Vaucouleurs cannot both be right. They may both be wrong. But there are some quite separate, powerful arguments in favour of the smaller value of H, and larger age, for the Universe.

THE AGE OF THE GALAXY

The first estimates of H gave an "age of the Universe" which was less than the age of the Earth, deduced by geologists. This conflict provided a powerful incentive for astronomers to find out what was wrong with their estimate of the age of the Universe, since the Universe itself must, clearly, be older than any star or planet in the Universe. The present range of estimates for H gives a range of possible "ages of the Universe" which are ample to accommodate the known lifetime of the Sun and Solar System, which is now thought to be about $4\frac{1}{2}$ billion years. But there are much older stars and star systems in our Galaxy, and the oldest of these have been around for long enough to rule out straightforward versions of the cosmological models with large values of H and with as much matter as there now seems to be in the real Universe.

Astronomers believe that they have a good understanding of how a star works. Their understanding of the processes of nuclear fusion inside stars helps them to understand the nature of the HR diagram, the relationship between the colour of a star and its brightness that is so useful in helping to determine distances inside our Galaxy. But the diagonal band of bright stars in the HR diagram corresponds to stars like our Sun, which are young enough to be "burning" hydrogen into helium in their hearts. Stars with different masses, but all busily burning hydrogen, sit along the main sequence band of the HR diagram. When their hydrogen fuel is exhausted, however, their appearances change, in a way that can be thoroughly explained by computer models of how stars work, and which can be understood in outline from a few simple physical arguments.

In the heart of an ageing star, at the end of its life as a member of the main sequence, there lies a core of helium surrounded by a shell in which hydrogen is still being converted into helium. The shell spreads outwards, and the core gets bigger, as the star ages. The helium core itself contracts under its own weight and heats up, until eventually it becomes hot enough at the centre for a new phase of nuclear burning to begin, with helium now being converted into carbon. This will happen to our Sun in about five billion years from now. With a small, hot core pouring out even more energy than the Sun does today, the effect on the outer layers of the star will be to make it swell up, engulfing Mercury, Venus and the Earth itself. The temperature at the surface of this huge ball of gas will be much less than that of the surface of the Sun today, so that the star will have a cool, red colour – such a star is known as a red giant, and many red giants are known to astronomers.*

These changes can be seen taking place in the stars of a cluster, when their brightnesses and colours are plotted in an HR diagram. The main sequence runs diagonally across the diagram, from top left to bottom right. Red giants lie above the main sequence, in the top right of the diagram. And although these changes take too long for an individual star to be observed changing its position across the dia-

---

* *More* energy escapes from the red giant, even though its surface is cooler than our Sun's surface, because the surface area is so much bigger. The amount of energy crossing each square metre is less, but there are many, many more square metres each contributing their share. A red giant actually emits a hundred times as much energy as our Sun does today.

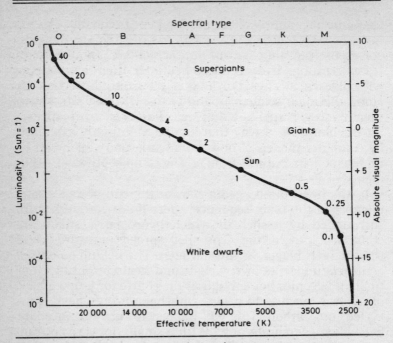

*Figure 5.2/*Concentrating on the main sequence, we can
see how the position of our Sun compares with those
of other stars. Luminosities are given in terms of the
brightness of the Sun; numbers along the main
sequence are the masses of the appropriate stars in
terms of the mass of the Sun.

gram, the computer models tell us exactly how the red
giants got there.

Stars with more mass burn their nuclear fuel more
quickly, and shine more brightly, than stars with lower
mass. They have to, simply to hold themselves up against
the inward tug of gravity. Such high mass stars lie on the
main sequence at the upper left of the HR diagram. As
their hydrogen fuel is exhausted, they "move" upwards and
to the right, off the main sequence. And as time passes all
the stars in the main sequence peel away to the right,
starting with those at the top left and finishing with those
at the bottom right. When astronomers study clusters of
stars in our galaxy, this is exactly what they see – a main

sequence starting out happily enough from bottom right, with low mass stars, but ending at some point and turning off to the right, into the red giant branch. If we know the distance to a particular cluster, we can compare the stage of evolution it has reached against the standard computer models simply by measuring the point where this turnoff occurs, and that immediately gives an estimate for the age of the cluster.

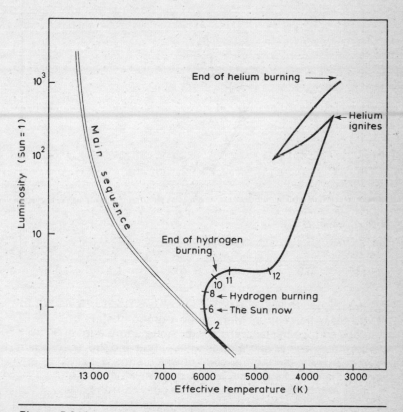

Figure 5.3/ As a star like our Sun ages, it leaves the main sequence and becomes larger but cooler. Numbers along the solid line indicate the age of a one solar mass star in billions of years since its formation; our Sun has, in fact, already begun to leave the main sequence and is four or five billion years old.

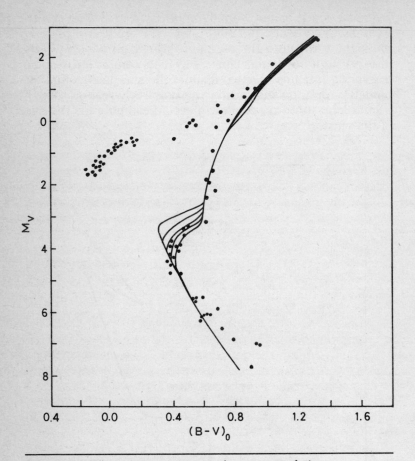

Figure 5.4 / When astronomers study a group of stars
that formed together and are all about the same age (a
cluster), they can obtain a good idea of how old the
stars are by measuring the point at which stars are
leaving the main sequence. Massive stars, which are
brighter and hotter than less massive ones, are the first
to turn off in this way. In this case, the cluster is M92
and the five curves modelling the turnoff represent
calculated ages of 10, 12, 14, 16, and 18 billion years.
The age of the cluster is interpreted to be between
14 and 16 billion years. Only about ten clusters have
well-determined ages, but these estimates are crucial
to an understanding of the age of the Universe.

As always in astronomy, there are uncertainties in the practical application of the technique. Corrections have to be made for the effects of interstellar dust on the light from stars as it crosses space to us; the turnoff point on the main sequence is never as precisely defined as I have made it sound; and there are other difficulties. With all these uncertainties, however, it is clear that the ages of the oldest clusters of stars in our Galaxy lie in the range from 14 billion to 20 billion years, always assuming that the standard models of stellar evolution are indeed correct, with a best estimate of 16 billion years.

There are other techniques which give similar ages for objects within our Galaxy. Radioactive isotopes found on Earth and in meteorites, for example, are thought to have been produced by supernova explosions within our Galaxy. These radioactive isotopes are unstable, and decay into stable elements in accordance with very precise rules that are well known from laboratory studies. The proportion of different radionuclides, as they are called, made in supernova explosions can be calculated using the same techniques which have proved so successful in explaining how stars work, and how they manufacture heavier elements out of hydrogen and helium. So the proportion of each kind of radioactive material left in the Solar System today can be used to determine the age of the Galaxy, assuming supernovae were manufacturing radionuclides at a steady rate from the time the Galaxy formed until the Solar System was born. That gives an age of about 15 billion years, in fair agreement with the other evidence.

This is not quite enough to rule out de Vaucouleurs' model with H equal to 100. In that model, the maximum age of the universe is just 10 billion years, and that would correspond to an almost empty universe, with very little matter around to slow down the expansion. The difference between 10 billion and 15 billion is not enough to settle the argument, given all the uncertainties involved. But the discrepancy certainly looks uncomfortable if there is enough matter around to close the Universe, when, if $H = 100$, the implied age is only 6.5 billion years, less than half the age of the oldest known stars. Sandage's estimate, $H = 50$, implies an age of 13 billion years if the Universe is just closed, close enough to the estimates of stellar ages to make us feel more comfortable. At a meeting of the Royal Society in March 1982, where element creation was discussed in the context of the big bang, Roger Tayler, of the University of Sussex, summed up the age "problem" by saying that "if omega equals one, it is necessary that H_0 is about 50 kilometres per second per Megaparsec", and nothing has happened since 1982 to change that interpretation of the evidence. The agreement is better still if H is just 40 kilometres per second per Megaparsec, when the age of a closed Universe rises above 16 billion years – and such a value of H certainly cannot be ruled out, using Sandage's figures. The other way to get near perfect agreement, of course, would be if the oldest star clusters in our Galaxy were actually a little younger than present estimates suggest. Could those estimates of stellar ages be just a *little* too high? Maybe, this time, the cosmology is telling the astrophysicists how best to refine their theories!

All these arguments would be resolved if cosmologists were able to measure the rate at which expansion of the Universe is slowing down today. That would tell us, once and for all, just how much matter the Universe contains, whether it is open or closed, and which value of the Hubble parameter is closer to the truth. Unfortunately, although such measurements are possible in principle, it seems unlikely that they can be achieved in practice in the immediate future.

THE REDSHIFT TEST

Hubble's law, the foundation stone of cosmology, is actually an imperfect description of the expanding Universe. That might seem like a flaw; but cosmologists are cunning enough to be able to turn the deviations from the straightforward Hubble law to their advantage, if only they could get enough good observations of very distant objects.

Unfortunately, the observations, as yet, are not up to the task. This is because the law that velocity (redshift) is proportional to distance *is* so good in our immediate neighbourhood, and almost as far out as we can see across the Universe. It is where the law starts to need amending that things get interesting, but that only happens so far from home that we cannot yet be quite sure just which amendments are necessary.

The reason things get interesting has to do with the geometry of space. On the surface of the Earth, an architect designing the floor plan of a building can happily use the geometrical rules laid down long ago by Euclid, which strictly speaking apply only to a flat plane, and doesn't have to worry about the curvature of the Earth. We all learned Euclidean geometry in school, and remember that the angles of a triangle, for example, always add up to 180°. But, as I mentioned in Chapter Three, if a team of surveyors were to lay out a series of very large, perfect triangles on the surface of a great, flat desert (perhaps somewhere in the Sahara) and then carefully measured the angles of those triangles, they would find that the angles always added up to slightly more than 180°, and that the difference from 180° was bigger for bigger triangles. This is because the surface of the Earth is actually curved, forming a closed surface that is very nearly spherical, and the "correct" geometry to apply in such cases is not the geometry of Euclid. In Chapter Three, we were interested in how remarkably flat the Universe is; now, it is time to see what the tiny deviations from flatness can tell us about its probable fate.

If space itself is curved, then the deviations of its geometry from the everyday Euclidean geometry we know so well will also show up over suitably large distances. "Suitably large", in this case, means over distances in excess of a few Megaparsecs – 10 million light years or more. This in itself is a significant feature of our Universe. It tells us that the geometry of spacetime is very nearly flat, and this in turn implies that the density of matter in the Universe is close to the critical amount required to make it closed. In principle, we can tell just how close by doing the equivalent of measuring the angles of huge

triangles. In practice, however, we lack the skill to distinguish which side of the dividing line between the open
and closed possibilities the Universe sits.

Measurements of the deviation of the Hubble law from
the simple "redshift equals constant × distance" are difficult. This, after all, is the law that is used to calculate the
distances to remote galaxies, by measuring their redshift.
How else can we estimate the distances to far away
galaxies, in order to *compare* these with the redshifts, and
see just how far out across the Universe the Hubble law
actually works? If all galaxies were the same brightness,
there would be no problem. The relative distances of all
the galaxies could be worked out simply by ranking them
in order of their brightness on the night sky – the dimmer
they appeared, the further away we would know they must
be. In fact, the situation would be a little more complicated
than it seems at first sight, even if all galaxies were the
same brightness. Where Euclidean geometry applies, the
brightness of each galaxy falls off in proportion to one over
the square of its distance – a galaxy twice as far away
seems only one quarter as bright. This simple rule itself
has to be modified when the geometry is different, and
corrections ought, strictly speaking, to be calculated for
each type of cosmological model. But that is the least of
the problems cosmologists face when trying to apply this
test.

It is only too clear from studies of galaxies that they are
not all the same brightness. And it is doubtful whether the
trick can be applied at all to the objects with the largest
redshifts, the quasars. Quasars are thought to be the highly
active, bright cores of galaxies. They shine even brighter
than ordinary galaxies and are visible further away, at
high redshifts, where the geometric effects should be
clearer. But there is no evidence that quasars all have the
same brightness, so the trick cannot be worked. One team
of astronomers, working at the Lick Observatory in California, has tried to use the brightness test on quasars
which have similar spectra to one another, and might therefore be expected to have similar brightnesses. For what it
is worth, this comparison suggests that the Universe is
closed, and will one day recollapse. Nobody, however, is

prepared to take this rather speculative interpretation of the quasar evidence at face value. Cosmologists are forced to restrict the technique to studies of galaxies, which are understood much better than quasars are, and where there is some hope of making the trick work.

Allan Sandage and his colleagues have carried out a long and patient study of galaxies and found some which do seem to have the same brightness as each other, and can be used as standard candles. Galaxies come in clusters, and very often the brightest galaxy in a cluster is a very large elliptical (so-called because of its shape, like a fat cigar; our own Milky Way Galaxy is a spiral, like the surface of a whirlpool, or the pattern made by cream stirred into a cup of coffee). As far as the observations can show, in the region of space out to a few Megaparsecs where the Hubble law holds precisely, the brightest large elliptical galaxy in any cluster is much the same brightness as the brightest large elliptical in any other cluster. There seems to be some natural maximum brightness such a galaxy can have, and any decent-sized cluster will have one galaxy that reaches the brightness limit. So, by using only these particular bright galaxies and plotting *apparent* brightness (in effect, distance) against redshift Sandage is able to see how far the resulting plot deviates from a straight line, and thereby determine how rapidly the expansion of the Universe is slowing down.

There are still problems. Remember that when we look at light from a galaxy 10 million light years away we see the galaxy as it was 10 million years ago. Can we be sure that the brightness of the galaxy has not changed during that time, as the galaxy and the Universe have evolved? Over this sort of range, there probably has not been much change. But for more distant galaxies, which we see in their youth, there is every likelihood of change. Astronomers would like to allow for this in the calculations, but have no independent means of telling how much brighter (or, conceivably, dimmer) galaxies were when the Universe was young. Some of the experts try to make educated guesses and allow for this luminosity evolution; others prefer to leave the observations alone, since any correction

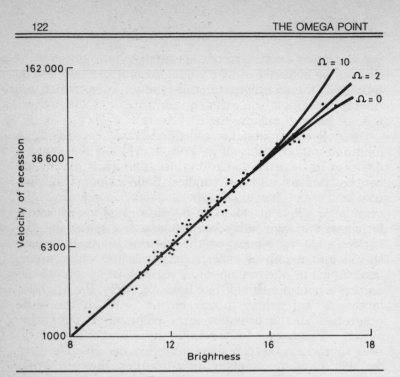

*Figure 5.5/*The Redshift Test
Because we see more distant galaxies as they were
when the Universe was younger, a comparison of their
redshifts with those of galaxies that are nearer to
us, and therefore look brighter on the sky, would in
principle reveal how quickly the expansion of the
Universe is slowing down. Unfortunately, in practice
these observations can only indicate that the Universe
is very nearly flat. They can only tell us that the value
of omega is roughly between 0 and 2. (All the calculated
curves overlap for small values of redshift and
brightness.)

they try to make might very well be in the wrong direction.

Having picked their way through this observational
minefield, cosmologists then have to compare the observations with their theoretical models. They calculate the
deviations from the simple Hubble law in terms of a deceleration parameter, often labelled q, which is defined in
such a way that $q = 1/2$ corresponds to $\Omega = 1$. At one

time, Sandages's plots of redshift against brightness seemed to favour a value of q of about 1, implying that there might be twice as much matter in the Universe as the minimum needed for closure; his latest plots, however, with more data gathered over the years, suggest that this early assessment was over-optimistic. Today, the best that this technique can do is to tell us that q probably lies somewhere in the range from 0 to 2, and that on these grounds alone the open Universe models cannot yet be ruled out.

Just after I had finished drafting this chapter, however, a new survey of galaxy redshifts was published in the *Astrophysical Journal* (volume 307, page L1). Edwin Loh and Earl Spillar, of Princeton University, reported a study of 1,000 galaxies that suggested the density parameter Ω is indistinguishably close to one. It is too early yet for this work to be regarded as definitive, but simply because it does depend on "old-fashioned" astronomy with optical telescopes it carries a powerful message to the observers that perhaps they should have been paying more heed to the theorists, who have long been pressing the case for a closed Universe.

The new test is one of several that can, in principle, be based on the geometry of the Universe.* If galaxies (or clusters) are distributed uniformly through the Universe, and geometry is Euclidean, then the number of galaxies we ought to see at different redshifts (different distances)

---

* Including one which uses the curvature of space, but doesn't involve any dynamics. Astronomers have discovered a handful of objects where the light from a distant quasar is bent around a galaxy in the line of sight between us and the quasar, because of the way the gravity of the galaxy distorts spacetime in its vicinity. The effect produces a double or triple image of the quasar as seen from Earth, and because light takes longer to reach us round one side of the intervening galaxy than the other, when one image changes its brightness, or flickers, the other may not do so for years, and then flickers in exactly the same way as the equivalent image of the quasar reaches us by the other route. By comparing the way these images change, and measuring the time delay, cosmologists can calculate how long it takes for the light to reach us from the quasar, and thereby deduce the distance to it independently of any redshift measurement. The first (and so far only) application of this technique gave a value for H of 75 km/sec/Mpc – smack in between the values favoured by Sandage and by de Vaucouleurs. A lot more measurements of this kind will be needed, however, before the new technique is given much weight.

can be calculated by the rules of geometry we learned at school. Broadly speaking, equal volumes of space ought to contain equal numbers of galaxies. If the geometry is non-Euclidean, however, when we try to determine equal volumes calculated in accordance with Euclid's rules, and then count the numbers of galaxies in them, there will be a discrepancy. "Equal volumes" far from us will contain either more or less galaxies than equivalent Euclidean "equal volumes" closer to us, and the exact number of galaxies more or less than the number predicted by Euclidean geometry would tell us the fate of the Universe. In particular, if $\Omega = 1$ and the Universe is not expanding, the geometry is Euclidean; and if $\Omega = 1$ and the Universe *is* expanding, we will see a certain, precise deviation of the "number counts" from the Euclidean prediction, because galaxies have moved apart as the Universe evolves.

Such deviations in the number counts are observed for many kinds of astronomical object – including radio galaxies and quasars – but it has proved very difficult to interpret the counts unambiguously. All of the studies of the large-scale dynamics of the Universe tell us that the geometry is very nearly Euclidean, that the expansion of the Universe is indeed slowing down, and that the amount of matter in the Universe must be fairly close to the amount needed for closure; but only the new study by Loh and Spillar tells us just how close to the dividing line we sit.

The Princeton team used the latest sensitive detectors to look at five tiny patches of the sky, each measuring about 7 arc minutes by 10 arc minutes (the Moon is 30 arc minutes across), and measured the redshifts of every galaxy they could detect in each patch. Because redshift is a distance indicator, they were, in effect, counting the number of galaxies in each of five pyramid shaped volumes stretching out into the Universe from our Milky Way. By comparing the number of low redshift galaxies with the number of high redshift galaxies they could determine the geometry of each pyramid, without getting embroiled in the problems of the absolute brightness of each galaxy. Each of the five pyramids contained about 200 detectable galaxies, and the survey stretched out to distances of about a thousand Megaparsecs – or, put another way, they were looking back in time about one-fifth of the way to the big bang itself.

Just how you interpret these number counts depends on exactly what kind of mathematical cosmological model you choose. Loh and Spillar chose the simplest, a version of the equations of relativity developed by Einstein and Willem de Sitter in 1932. This is particularly important, since it rejects an earlier idea that Einstein brought in to the equations, and later described as the biggest blunder of his career.

When Einstein first solved the equations of general relativity to describe the behaviour of spacetime containing matter (that is, the Universe) he found that the solutions all described dynamic models, including the case of universal expansion. At that time, seventy years ago, astronomers thought that the Universe was static, and that the Milky Way, *was* the Universe. Einstein had to introduce a new term, the cosmological constant, to hold his models still. But within a few years the discovery of the cosmological redshift and the expansion of the Universe beyond the Milky Way removed the justification for the cosmological constant.

That has not stopped some cosmologists from tinkering with solutions to the equations that include all kinds of cosmological constants, which can make models expand faster or slower to fit the pet theories of the mathematicians. It's a kind of game they play, with little relevance to the real world. Purists, however, have always argued that because the constant is not necessary, it should not be introduced at all, or should be set equal to zero – the position Einstein subscribed to once the redshift effect had been identified. Very recently, Stephen Hawking has developed an elegant mathematical proof that although the cosmological constant *might* not be zero, there is an overwhelming likelihood that it *will* be zero (rather like the argument that although all the air in your room *might* rush off into the corners, there is very, very little chance that you'll ever see it happen!). Now, there is a weight of observational evidence to back the purists up, as well –

ample justification for cosmology's embarrassing constant to be ignored in the rest of this book. If the simplest solution to Einstein's equations works, why frighten ourselves with unnecessary complications?

Loh and Spillar found that their counts of galaxies at different redshifts can best be explained by the simplest Einstein-de Sitter model, with zero cosmological constant and a value of omega of 0.9, with technical "error bars" of ± 0.3 – indistinguishably close to one. It doesn't matter what the material holding the Universe together is, or even whether it is clustered in clumps the way galaxies are. It just has to be there.

This discovery, using the traditional technique of redshift measurements, caused a great deal of pleasure to one group of astronomers who had been tackling the same problem from a very different angle. They, too, had reached the conclusion that $\Omega = 1$, and had presented their conclusions a few months before the work of Loh and Spillar appeared. Those conclusions had been regarded with deep suspicion by some astronomers unused to the new tools of infrared astronomy from satellites. But this dynamic measurement of the way our own Galaxy is moving through space, which comes down unambiguously on the side of the closed models, can scarcely be dismissed as unreliable when the traditional techniques are now sending us the same message.

THE ATTRACTION OF VIRGO

Before we can have any confidence in estimates of the distance scale of the Universe based on studies of the redshift-distance law, we really need to be sure we understand all of the Earth's motion through space. Our home planet, on which all our telescopes are based, is moving around the Sun; the Sun is moving around the Galaxy; and the Galaxy is moving with respect to other nearby galaxies. Although all galaxies – or, rather, all *clusters* of galaxies – are being moved apart from one another by the universal expansion, they can each have quite substantial "peculiar" velocities of their own, as they orbit around one another. M31, for example, is actually moving *towards* us, at present, because the galaxies in the Local Group are held together as a group by gravity, and are not being steadily pulled away from one another by the universal expansion. On this scale, the local gravity overwhelms the expansion effect. Cosmologists need to know all such local effects, and subtract them out of their calculations, before they can be sure that they are left with the pure redshifts due to the expansion of the Universe. We need a fixed frame of reference, a stationary platform which only moves with the universal expansion. Without such a reference point, all of the redshift surveys are at least partly based on guesswork.

The problem has been highlighted by changing opinions on just how big an effect the Virgo Cluster has on the motion of our Galaxy and the Local Group. We certainly are moving away from the Virgo Cluster, relatively speaking; the redshift shows that. But we would expect that the gravitational influence of all the matter in the Virgo Cluster would be holding us back a little, impeding our relative recession from the cluster caused by the expansion of the Universe. Confusingly, astronomers sometimes refer to this virgocentric pull in terms of a velocity of the Milky Way towards the Virgo Cluster, as "infall"; what they mean is that we are receding from the Virgo cluster that much slower than we might expect from the Hubble law alone. But how much slower?

We can get some idea of what is going on by looking at the redshifts and distances of the galaxies in our immediate vicinity, where some astronomical techniques still provide an indication of distance independent of the redshift. If we look out in opposite directions in space and find two galaxies at roughly the same distance from us, but one of them has a slightly larger redshift than the other, we know that the excess redshift must be due to the peculiar motion of the Milky Way, or of one of the other two galaxies. If we look at enough galaxies in this way, we can hope that all the odd motions of the other galaxies cancel out, and any consistent tendency for redshifts on one part of the sky to be lower than those on the opposite part of the sky is a sign that our Galaxy has a peculiar velocity towards the low redshift region. This kind of technique has been applied to try to determine the size of our infall "towards" the Virgo Cluster, but with only limited success. Different astronomers have come up with different estimates of the size of the effect, ranging from virtually no infall up to about 500 kilometres per second. A lot depends on which groups of galaxies are used to calibrate the motion of the Milky Way – and the astronomers who do the calculations are uncomfortably aware that their figures will still be distorted if all the galaxies being used in the calibration are themselves being held back by the Virgo Cluster in the same way. No peculiar motion of our Galaxy compared with the expansion of the Universe will show up if we try to measure the effect by comparing the motion of the Milky Way with the motion of a lot of other galaxies all streaming in the same direction we are.

Nevertheless, studies of the virgocentric flow are beginning to tell us something about the distribution of matter in the Universe.

The Virgo Cluster is just close enough for some estimates of its distance to be made using a variety of different secondary techniques. These involve some subtle astronomical reasoning, and do not all give the same "answer" – indeed, the same technique applied by two different astronomers will often give two different values for the distance. The range of estimates is from about 16 Mpc to 22 Mpc, with 20 Mpc being a reasonable compromise between the extremes. Because of the uncertainty in the estimates of the peculiar velocities of the Milky Way itself and of the galaxies in the Virgo Cluster, this cannot be used directly to determine H. Instead, by comparing the brightnesses of individual galaxies within the Virgo cluster and supernovae within those galaxies with their counterparts in a much more distant group of galaxies, the Coma Cluster, astronomers deduce that the Coma Cluster is about six times further away than the Virgo Cluster –

that is, some 120 Mpc. The Coma Cluster is so far away that it has a redshift corresponding to a velocity of 7,000 kilometres per second, far greater than the few hundred kilometres per second of the Milky Way's peculiar motion. So, at last, we have a more or less direct comparison of distance and redshift on a scale big enough to be sure that the peculiar motion of the Milky Way cannot be introducing an error of more than about 10 per cent. The long chain of reasoning yields a value for H of between 45 and 55 km/sec/Mpc. But this isn't the end of the Virgo story.

The strength of the pull which the Virgo Cluster exerts on the Milky Way depends on how much matter there is in the cluster. With this value of H, astronomers know how big the redshift "ought" to be, and by comparing this with the measured redshift they deduce that the effect of the attraction of Virgo is equivalent to a motion towards the Virgo Cluster at a speed of a little over 200 km/sec. The amount of matter needed in the Virgo cluster to produce this effect is equivalent to a density of about one tenth of the value required to close the Universe. Even if the "infall" velocity is as high as 450 km/sec, we still only "need" enough matter in the Virgo cluster to produce a value for omega of about 0.25, if the same density of matter were spread uniformly throughout the Universe.

This would be a very powerful argument in favour of the Universe being open – if we could be sure that all of the matter in the Universe is distributed in the same way the bright galaxies are distributed (and assuming, of course, that the Virgo Cluster is typical of the Universe at large). But if there is any independent evidence of the Universe being closed, then what the attraction of Virgo is telling us is that not only is most of the matter in the Universe not in the form of bright stars, it isn't even distributed throughout the Universe in the same way that bright stars and galaxies we can see are distributed. What we need is a way to measure the distribution of matter over even greater volumes of space, looking at radiation in wavebands different from those of visible light on which astronomers have depended for so long. Ten years ago, it would have been an astronomical pipe-dream. In 1986, it became reality.

MICROWAVES AND THE MOVEMENT OF THE MILKY WAY

The technique of measuring the peculiar motion of our Galaxy through space – independent of the expansion of space itself – works best, in principle, using the redshifts of more distant galaxies. But the further away galaxies are, the harder it is to estimate their distances, and the less confidence we can have in the accuracy of the calculation. Nevertheless, back in 1976 Vera Rubin and her colleagues, of the Carnegie Institution of Washington, tried to extend the technique by comparing the motion of our Galaxy against the "frame of reference" provided by a spherical shell of distant spiral galaxies. These galaxies are all at a distance of about 100 Megaparsecs from us, assuming the Hubble parameter really is close to 50 in the usual units. They surround us in the way that the skin of an apple surrounds a pip at its centre, and they are so far away that it is reasonable to expect that all their own little, peculiar motions will average out, and that taken together they provide a reference-frame moving only with the expanding Universe. Rubin's calculations showed that, relative to these distant galaxies, our Milky Way (and the Local Group) is moving through space with a velocity much bigger than anyone had expected – 600 kilometres a second, over and above our motion as part of the universal expansion. The discovery was so surprising, and the velocity revealed by the technique so large, that most astronomers simply refused to accept it. They could just about cope with an "infall" of 200 or 300 km/sec towards the Virgo Cluster, where they could see evidence, in the form of bright galaxies, of the matter doing the pulling. But a velocity of 600 km/sec, in the direction of nothing in particular on the night sky, where there was no bright cluster of galaxies visible? Ridiculous!

Ten years later, the notion no longer seemed so ridiculous, and Rubin was vindicated. Two new pieces of evidence combined to change the opinion of the astronomers.

The first insight came from studies of the microwave background radiation, the leftover hiss of radio noise from the big bang itself. This radiation has filled the Universe since very shortly after the moment of creation, but it has not been affected by the material content of the Universe since electrons combined with the nuclei created in the fireball to make electrically neutral atoms. This kind of radiation can only interact with free charged particles; but within a million years of the moment of creation all of the positively charged protons and negatively charged electrons were locked up in neutral atoms of hydrogen and helium. Ever since, the background radiation has simply expanded with the Universe, cooling and weakening as it is redshifted to longer and longer wavelengths, but never being disturbed by matter. The background radiation ought to provide the best available frame of reference in the expanding Universe, an ideal basis against which to compare our own peculiar motion. And it does.

As observations of the background radiation have improved over the past twenty years, astronomers have been able to go beyond merely noting its existence and taking its temperature (observations that were instrumental in establishing the big bang description of the Universe) and have mapped the strength of the radiation over almost the entire sky, at many different wavelengths, using instruments that are now sensitive enough to measure small differences in the strength of the radiation – small temperature differences – from different parts of the sky. The techniques involve observations from the ground, from high-flying aircraft, from balloons which carry instruments above the bulk of the atmosphere, and from satellites in orbit around the Earth. By the middle of the 1980s they were showing, unambiguously, that there is a warm patch in the cosmic background, in a direction roughly at 45 degrees to the direction of the Virgo Cluster, and a cold patch in the opposite direction on the sky. The warm patch is equivalent to a region of blueshifted background radiation, where the wavelength has been shortened slightly because we are moving towards the incoming waves; the cold patch is a region of redshift, caused by our motion away from the incoming waves. The interpretation of the discovery is clear – we are indeed moving at a high velocity relative to the background radiation, and therefore relative to the overall expansion of the Universe. It is exactly the 600 km/sec velocity Rubin found ten years previously.

At first, some astronomers thought that this motion might be due to the gravitational pull of a concentration of matter in a group of galaxies known as the Hydra-Centaurus Supercluster. If the Milky Way were being tugged one way by the Virgo Cluster and another way by the Hydra-Centaurus Supercluster, the overall effect could be to produce a movement towards a point in space roughly midway between the directions to the two masses. But that notion seems to have been squashed by very recent discoveries, announced only in 1986. A massive study by a team of astronomers from six different institutions around the world, from Herstmonceux in Sussex to Pasadena in California, reported their investigation of the motion of 400 elliptical galaxies, spread evenly across the sky, to an international meeting in Hawaii. Using chains of argument like the one which seems to work so well for the Coma Cluster, they were able to work out distances and peculiar velocities for all these galaxies. They found that *all* of the nearby galaxies and groups of galaxies are being tugged through space in the same way as our Galaxy and the Local Group. The Virgo Cluster, the Hydra-Centaurus Supercluster, the Local Group and others are all moving, at 600 to 700 km/sec, towards a region *beyond* the Hydra-Centaurus Supercluster.

Where does the region involved in this streaming of galaxies end? And just how much matter do you need to pull so many galaxies so strongly? The best answers to these questions are provided by surveys of distant galaxies carried out by the Infrared Astronomy Satellite, IRAS, and, once again, reported in the mid-1980s.

WEIGHING THE INFRARED EVIDENCE

All studies of the distribution of galaxies seen in visible light are handicapped by a phenomenon known as reddening. This has nothing to do with the redshift, but is a dimming and reddening of the light from distant objects caused by dust in the Milky Way itself – the effect is exactly equivalent to the way dust in the atmosphere of the Earth causes red sunsets. The dust of the Milky Way simply blocks out light from many regions of the sky, leaving astronomers with reasonably clear views of only parts of the northern and southern hemispheres of the sky, above the plane of the Milky Way. Light from faint galaxies (and that means, by and large, more distant galaxies) is affected worse, so the further out into the Universe you want to look the higher you have to raise your astronomical sights, to high latitudes in the northern or southern skies. Then there is the problem of comparing the northern galaxies with the southern ones. When they try to combine the limited observations they do have to provide a catalogue of galaxies covering as much of the sky as possible, the astronomers find it impossible to assess the brightnesses of northern and southern galaxies precisely in one definitive scale. Northern galaxies can only be studied by telescopes in the northern hemisphere; galaxies high in the southern sky are only visible to southern telescopes. Ideally, comparing the brightnesses of faint objects, measured at the limit of present-day techniques, to the precision required by these dynamic studies, requires that all the galaxies being studied are monitored with the same combination of telescope and instruments. But there is no way to use the same telescope and instruments to measure the brightness of every galaxy visible from the surface of the Earth; the telescopes are simply too unwieldy to move about.

IRAS solved both these problems, and others. Infrared light is scarcely affected by reddening caused by dust in the Galaxy, and the same instruments, in orbit around the

Earth, were used to map the entire sky. IRAS could see galaxies in all directions except for a very narrow region of the sky across the Milky Way itself, and these galaxies could easily be distinguished from bright stars in our own Galaxy. The result was a survey of tens of thousands of galaxies at infrared wavelengths, covering almost the whole of the sky.

Some bright infrared galaxies have also been identified using optical telescopes, and their redshifts measured. By comparing the brightnesses of these objects in the infrared with those of other infrared galaxies that have not yet been studied optically, it seems that the IRAS survey extends out to distances at least twice as great as those of the galaxies studied by Rubin and her colleagues. But they are not distributed uniformly across the sky. On average, there are slightly more sources in same sized areas of sky on one side of the sky than the other, and the direction picked out by the IRAS survey is almost exactly the direction in which we are moving compared with the cosmic background radiation. At last, astronomers can actually "see" (using infrared detectors) evidence of a concentration of matter in the right direction to be producing the pull affecting the Local Group and other galaxies in our part of the Universe.

This isn't the end of the story. Michael Rowan-Robinson, of Queen Mary College in London, is one of the researchers involved in analysing the IRAS data. He has calculated how much matter there would have to be overall, distributed across the region of the Universe surveyed by IRAS in the same way that the IRAS galaxies themselves are distributed, in order for the extra concentration in the direction we are moving to produce a gravitational tug strong enough to give the Local Group a peculiar velocity of 600 km/sec. The answer he comes up with, and which he announced to the Royal Society meeting in November 1985, is exactly equivalent, within the inherent uncertainties of the approximations involved, to the density required to close the Universe. The simplest interpretation of the IRAS data is that omega is almost exactly one, and that the distribution of galaxies on the night sky picked out by the limitations of ground-based observations of *visible* light is *not* a good guide to the way matter is distributed across our region of the Universe.

This is the most powerful piece of evidence concerning the nature of the Universe that studies of the dynamics of galaxies can yet provide. It is the first direct measurement of galaxy dynamics that gives a value for omega of one, and it holds good even if the evidence for large-scale streaming of all the local clusters and superclusters, the latest and most controversial piece of evidence, does not stand up to further scrutiny. These are new ideas, and they will undoubtedly be modified in the months and years ahead. But the basics seem clear, and they mesh in very nicely with the requirement that $\Omega = 1$ from inflationary models of cosmology and from the need for the Universe to recollapse in order to preserve the electromagnetic arrow of time. Remember, too, that while it is always possible that we may find more matter, and more kinds of matter, in the Universe than we yet know about, there is no way that we can ever "remove" the matter we have already found. That is an absolute bottom limit on the density parameter; estimates of omega must always go up as time passes, never down. Most astronomers are still cautious about making any dogmatic claims, but the evidence in favour of a closed Universe is better than it has ever been. Even if the Universe is not closed, there must still be much more matter out there than is allowed for by the production of baryons in the big bang. Non-baryonic matter definitely dominates the Universe, and probably makes it closed. And although the way galaxies move through the Universe cannot tell us what and where that matter is, the way in which galaxies are distributed in space, the pattern they make today in a picture frozen in one instant of cosmic time, does provide some strong clues in that direction.

CHAPTER SIX

CLOSE TO CRITICAL

Dark matter dominates the Universe. At least 90 per cent of everything, and perhaps 99 per cent, has never been seen. The bright stars and galaxies are not even the tip of the iceberg of the material Universe, since at least all of an iceberg is made of ice, and is attached to the tip. We don't know what the dark matter is, or where it is; only that, unlike stars, it cannot be baryons and that there is enough of it to place the Universe close to the critical dividing line between being open and being closed. The search for the missing mass is now intensifying, with both observers and theorists making their contributions to the debate. The answers are not yet in, and it would be foolish, in a book being finished early in 1987, to pretend that the identity of the missing mass had been uncovered. But it is already possible to narrow the search down, to eliminate some of the candidates, and to get a good outline idea of what kind of particles it is that really dominate the Universe. Studying galaxies as they are today provides important clues.

THE HALO CONSPIRACY

First, *where* is the dark matter? The simplest guess would
be that it is all in galaxies, after all. Perhaps there is a lot
more material, in the form of dust, planets, black holes or
something even more exotic, than we see as bright stars.
In fact, to bring the Universe close to critical it would
mostly have to be something exotic, since all the other
possibilities are made of baryons, and the helium abundance sets such a tight limit on how much baryonic matter
there can be in the Universe. In recent years, observers
have found evidence that there *is* a great deal of dark
matter in spiral galaxies – in some cases, at least four times
as much dark matter as there is in the form of bright stars.
But this is far from being enough to push omega close
to 1; indeed, you can just about account for all of this
galactic dark matter within the limits allowed for baryons.

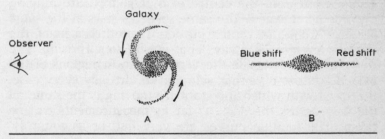

Figure 6.1 / When we view a spinning spiral galaxy edge
on, we can calculate how rapidly it is rotating by
measuring the redshift and blueshift at different
distances out from the central bulge.

These are new discoveries, made only in the past few
years. They depend on measuring the speed with which
spiral galaxies rotate. Of course, astronomers cannot watch
the changing pattern of stars as a galaxy turns slowly
about its nucleus; it takes hundreds of millions of years to
complete one turn. But when they look at spiral galaxies
which happen to be oriented edge on to us in space, so

that they are seen as thin disks, astronomers can use the Doppler shift technique to measure how fast the stars on one side of the disk are moving towards us, and how fast the stars on the other side are moving away, as the disk rotates. Modern spectroscopic techniques are so sensitive that they can measure these Doppler shifts at different distances out from the centre of the tiny image of a galaxy captured by a large telescope, and thereby give velocity measurements across its disk. In recent times these measurements have been extended further out, beyond the region of bright stars, to measure the velocities of clouds of hydrogen gas, still part of the disk of the distant galaxy, using radio astronomy techniques. The results were startling when they began to come in in the early 1980s.

When astronomers plot out the velocities of the stars and clouds orbiting in the disk of a distant galaxy at different distances from its nucleus, they obtain what they call a rotation curve. These curves are usually very symmetrical. The stars on one side of the distant galaxy, at a certain distance out from the centre of that galaxy, are moving towards us at exactly the same speed as stars at the same distance from the centre but on the other side of the nucleus are moving away. That wasn't so surprising. The surprise is that outside the very innermost regions of the galactic centre, on either side and in virtually every case, the speed with which the stars are moving is the same all the way across the disk, as far as measurements can be taken. The rotation curves are very flat, or as some astronomers quip, the most striking feature of rotation curves is that there are no striking features.

This came as a surprise because astronomers had assumed that the greatest amount of mass in a spiral galaxy was concentrated where the galaxy shone brightest, in the central nucleus where there are many stars. If that were the case, then stars further out from the centre should be moving more slowly in their orbits, in exactly the same way that the outer planets of our Solar System (Jupiter, Saturn, Uranus, Neptune and Pluto) move more slowly than the inner planets (Mercury, Venus, Earth and Mars). This is a straightforward consequence of the law of gravity discovered by Newton. There are only two ways to account for the flatness of the rotation curves of spiral galaxies. Either Newton's law is wrong (a possibility which has been actively considered by some researchers, but seems rather a drastic assumption) or there must be a great deal of dark matter spread in a huge, roughly spherical halo around each spiral galaxy, and dragging the bright stars round with it as it rotates. Most of the mass, in other words, *cannot* be associated with the bright stars in the central nucleus.

This is the dark matter that can – just – be explained within the allowance of baryons provided for by big bang cosmology and the helium abundance. Galaxies like our own Milky Way are embedded in vast haloes of matter,

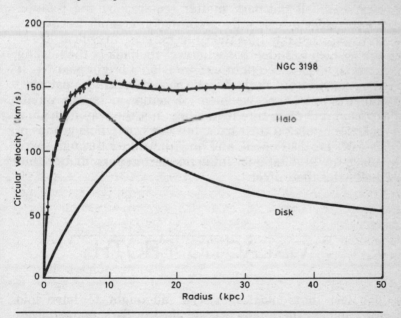

*Figure 6.2/*The actual rotation curves of galaxies such as NGC 3198 are very flat. This can only be explained if the amount of dark matter in the halo of the galaxy rises, as indicated, while the amount of matter in bright stars falls off outside the central bulge. The two effects "conspire" to produce the observed rotation curve. (Figure supplied by Tjeerd van Albada.)

which might be in the form of objects like large planets ("Jupiters") or failed, dim stars ("Brown Dwarfs"), or black holes (or it *might* be something more exotic). For some reason, which nobody yet understands, the amount of matter in the halo exactly balances the expected fall-off of rotational velocity from the centre of each galaxy, to produce the featureless rotation curves – a "conspiracy" that astronomers believe must be more than a coincidence, but which they are at a loss to account for.

Dark matter dominates galaxies, just as it dominates the Universe, and unravelling the way dark matter and bright stars interact in galaxies will keep many astronomers busy for many years. But this is only an incidental feature of the search for the mass needed to close the Universe, since even with all the dark matter required by the rotation curves galaxies can only account for a small fraction of the mass needed to take the Universe close to critical. These studies help to show that wherever the bulk of the missing mass is, it is *not* tied tightly to individual bright galaxies. It is somewhere out in the black space between the galaxies, and the only hope we have of getting a handle on its location and its nature is to study, not the way individual galaxies rotate on their axes, but the way whole groups of galaxies are distributed, and how they move through space under the gravitational influence of the dark matter that dominates the Universe.

GALACTIC FROTH

The fact that galaxies exist at all ought to have told astronomers there was a great deal of dark matter, even before the new wave of astronomical discoveries rammed the point home in the 1980s. In an expanding Universe, which starts out with a uniform distribution of matter (as required by the smoothness of the microwave background radiation), it is very difficult to make clumps as big as galaxies if the overall density is as low as the baryon limit requires. In the expanding Universe, things are being pulled apart and stretched thinner, not clumped together. So how do clumps as big as galaxies grow? There must be some deviations from perfect uniformity early on, some regions where there happens to be a little extra material, and others where there happens to be a little less. Once a region of extra density reaches a certain size it will continue to grow, as its gravity tugs in more matter from outside, holding it back against the general expansion. But if the Universe has only the density implied by the baryon limit, then by the time the original hot gas has cooled to the point where gravity can begin to hold clouds of gas together, it is already so thin that no realistically likely density fluctuation will be big enough to do the job of making a galaxy.

In the 1970s, there were two lines of attack on the problem of how galaxies formed. One view, espoused by Jim Peebles in the United States, held that things had grown from the bottom up. Without knowing how the first "seeds" formed and grew, he suggested that galaxies formed first, and that only afterwards did galaxies clump together to form clusters of galaxies. Since galaxies are very old – the oldest stars in our Galaxy are almost as old as the Universe itself according to our best models – this idea clearly makes sense. But it makes a prediction, that galaxies and clusters should be distributed randomly throughout the Universe, which is not borne out by modern observations.

In the 1970s, the alternative to Peebles' "bottom up" scenario was the "top down" idea, purveyed by Yakov Zel'dovich of the Soviet Union. He suggested that in the heat of the early Universe any small-scale fluctuations would have been wiped out, and only very large-scale inhomogeneities would survive as the Universe began to cool out of the big bang. On his picture, the original "seeds" of the structure we see today were not galaxy-sized, but would have been equivalent to superclusters of galaxies, containing a thousand times more mass than the galaxy-sized seeds of Peebles' scenario. When these seeds collapsed into flat pancakes, said Zel'dovich, galaxies would form around the edges of the pancakes, and at regions of extra high density where two pancakes crossed. He predicted that the Universe would be found to contain strings and chains of galaxies, strung out along filaments like beads on a wire, with great, empty spaces in between the filaments. This is much closer than Peebles' prediction to the way we do see galaxies distributed today, which gives a boost to the pancake theory. But since galaxies form late in this picture, it is hard to see how there has been time for stars as old as those of our Galaxy to have been born in a Universe as young as the big bang calculations imply.

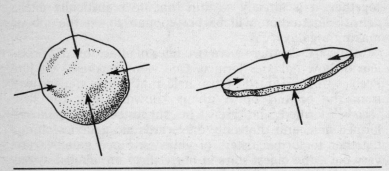

*Figure 6.3/*Pancake Theory
One view of the way galaxies formed involves the
collapse of huge clouds of material into flat pancakes.
Each pancake would then break up into many galaxies,
forming a cluster or supercluster.

Neither of the two rival theories of the 1970s really works, but they provided a jumping-off point for observers seeking to test the rival ideas, and then for theorists seeking to explain what the observers found using new ideas which go beyond those of the 1970s models.

The most striking feature of the new surveys of the Universe, three-dimensional maps of the distribution of galaxies based on redshift data, is that the pattern is full of holes. Careful mapping of galaxies with redshifts going up to the equivalent of distances beyond a thousand million light years shows that almost all galaxies congregate around the edges of great bubbles, up to 150 million light

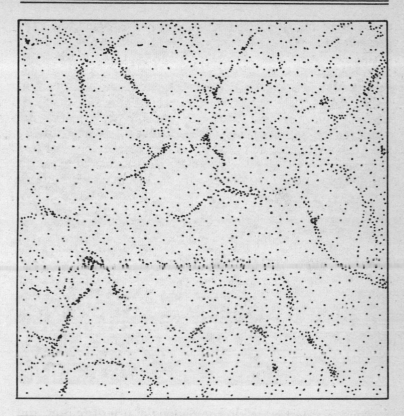

Figure 6.4 / Computer simulations of the pancake
process show how lines and chains of "galaxies" would
look on the sky, as we viewed the broken-up pancakes
edge on.

years across, which contain little or no luminous matter.
Observers see the same picture wherever they choose to
carry out their surveys; but unfortunately it is only possible
to survey in the necessary detail a very small part of the
sky. This has left plenty of scope for speculation about the
exact relationship between the voids and the galaxies,
puzzles which smack of hair-splitting at first sight but
which actually address fundamental questions about the
nature of the big bang and the location of the missing
mass.

The key question is, do the regions of bright matter, the chains and shells of galaxies, surround the voids, or do the voids surround the bright matter? Putting it in slightly more familiar terms, we can think of the pattern of galaxies on the dark sky as like a pattern of polka dots; but is it a pattern of white dots (galaxy clusters) on a black background, or one of black dots (voids) on a white background? In 1986, a team of American researchers tackled the problem by using a computer to study how the bright and dark regions of the Universe twine about one another – their topology. Richard Gott and Mark Dickinson at Princeton and Adrian Melott at the University of Kansas discovered that *neither* of the simple guesses is correct. The pattern is not one of black dots on a white background, nor is it one of white dots on a black background. Instead, the two structures are completely interconnected, like the

Figure 6.5/ (More than) A Million Galaxies
By combining the information from more than a
thousand photographic plates, astronomers have been
able to produce these images, showing more than a
million galaxies over the entire northern hemisphere
sky (left), and more than another million over a large
part of the southern sky (right). The overwhelming
visual impression is of a network of interconnected
chains and filaments of galaxies. How real are the
filaments, and what can they tell us about the
distribution of matter across the Universe? (From M.
Seldner, B. R. Siebers, E. J. Groth and P. J. E. Peebles,
Astronomical Journal, volume 82, page 249.)

structure of a natural sponge. Strictly speaking, there may
be only one "void" and one galaxy filament, twine
around each other in a complex fashion.

This topology explains many puzzling features of the observed pattern of galaxies on the sky, which is a two-dimensional projection of a complicated pattern in three dimensions, producing a messy appearance of interconnected clusters and voids mixing into one another. The discovery is very encouraging for the theorists, because it suggests that there is no real distinction between the regions of higher than average density where galaxies form and the regions of lower than average density, the voids. If the present-day structure of the Universe is simply a result of random fluctuations in the initial fireball, that is exactly what the simplest theories would predict – no preference for fluctuations that give a small region of high density *or* for fluctuations which produce similar regions of low density, but both kinds of fluctuation occurring at random. The Universe *looks* frothy, but is not, in fact, organised into regular cells, like the structure of a honeycomb.

Melott has taken the idea one step further, and has studied not just the "snapshot" picture of the distribution of galaxies today, but also the way they are moving in three dimensions. Astronomers can only detect movement of galaxies along the line of sight, the redshift effect; Melott has used computer models to create patterns of simulated "galaxies" moving in three dimensions, under the influence of the gravity of their neighbours, and has then converted these into the patterns that would be produced for velocities along the line of sight seen by a hypothetical observer riding on one of the galaxies. He finds that the right kind of pattern, to match observations of real galaxies in our Universe, only comes out if his simulated universes contain enough matter for omega to be close to one – but the one-dimensional view from any of the simulated galaxies always gives a false impression that the simulated universes contain less matter than this. It may be that at least some of the missing mass may be accounted for by a kind of cosmic optical illusion, which makes the standard redshift tests give us a false value for omega.

The frothy appearance of the Universe, like so many other factors, is pointing to a density close to critical; it is so very much in line with the idea that galaxies are a

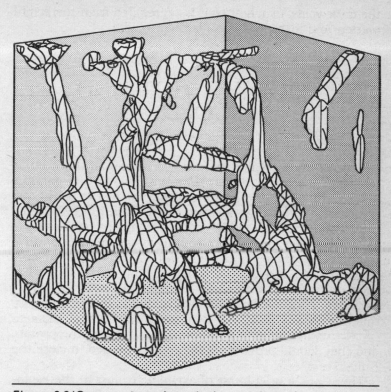

*Figure 6.6/*One way to make galaxies condense in
chains and filaments in the computer models is to put
in the influence of a background sea of low-mass
neutrinos, sufficient to close the Universe. This
computer simulation, provided by Adrian Melott, picks
out regions in which the density of "baryonic" matter
is twice as much as the average. The chains and
filaments are strikingly like those seen in the real
Universe. Although this particular model is no longer
the front runner, such a pattern cannot be produced
naturally in any of the simulations without the presence
of a great deal of dark matter.

result of small inhomogeneities dating back to the birth of
the Universe. The next step in the search for the missing
mass is to attempt to find out whether it is all associated
with the chains and filaments of galaxies that surround

the dark voids, or whether it is segregated from the bright matter, and lurks in the voids themselves.

BLOWING HOT AND COLD

When astronomers first became convinced of the need to find dark matter to account for the dynamics of galaxies and the Universe, the best candidate seemed to be the sea of neutrinos that they already knew filled space. At least the astronomers knew for sure that neutrinos existed – they are now a completely routine component of particle interactions studied both by theorists and experimenters using the big accelerators. The big bang models say that there must be an enormous number of neutrinos in the Universe, ghosting on their way without interacting very much with anything. All it needs is for each of them to have a tiny mass, one billionth of the mass of the proton, and they would provide all the matter needed to close the Universe.

There was a brief burst of excitement, just at the beginning of the 1980s, when two experiments in laboratories on the surface of the Earth came up with hints that neutrinos might indeed have just such a tiny mass. Since then, however, the tide has turned against neutrinos as the explanation for the dark matter. Those first hints of a possible neutrino mass, which require difficult and error-prone measurements, have not been borne out by further tests, and the issue of neutrino mass is still open, taxing experimenters around the world. Even more damningly, though, neutrinos do not, after all, seem to have the right properties to be able to account for the pattern of galaxies seen across the Universe.

In this connection, neutrinos are the archetypal example of the kind of dark matter that cosmologists label "hot". Individual neutrinos – and other hot particles – emerge from the big bang with very large velocities, close to the speed of light (relativistic velocities), and stream across the Universe freely in all directions. High velocity particles, free-streaming in this manner, tend to smooth out fluctuations in the density of other kinds of matter as the Universe cools – as Jack Burns, of the University of New Mexico, has graphically put it, this is "much as a cannonball moving at high speed might scatter a loosely built wall of stones without being appreciably slowed by the collision". In spite of their light mass, such hot neutrinos carry enough energy and momentum, because of their relativistic speed, to break up small-scale structures in the early Universe. This process would continue until the neutrinos cooled to the point where their speeds were, on average, about one-tenth the speed of light. From then on, density fluctuations could grow as the Universe continued to expand, producing a "top down" pancake universe rather like the earlier ideas of Zel'dovich.

Among other things, this simple hot dark matter scenario implies that there is little or no baryonic matter in the voids between bright galaxy superclusters, and that it has all been swept into the filaments and shells, or bubbles, of the froth. Some astronomers are now trying to probe the dark depths of the nearby voids, using the most sensitive telescopes and detectors to see if there are any galaxies to be found there. Already, there are hints that although the voids are much more sparsely populated than the froth, they do contain some dim galaxies, called dwarfs. That is bad news for all hot dark matter models, including the neutrino-dominated universe. There is also the seemingly insurmountable problem of the ages of stars in our own Galaxy, which don't allow enough time after the big bang itself for all the free-streaming and pancake collapse to take place before stars form. In round terms, galaxies must have formed only one or two billion years after the big bang, but computer simulations of the pancake process suggest that it needs up to four billion years to do the job. Sadly, even though it is such a straightforward and natural way to account for the large-scale distribution of bright matter across the Universe, astronomers have been forced to accept that at the very least hot dark matter is not the whole story, and that it may be a complete red herring.

The natural alternative to his scenario is to imagine that
the Universe is dominated by particles that move much
more slowly through space, and therefore do not destroy
small-scale fluctuations in density before they have a
chance to grow into galaxies. Such scenarios are called
"cold dark matter" models (CDM for short). The most
obvious problem with these models is that nobody has yet
detected any particle that could be a candidate for the
cold dark matter. But the astronomers were able to specify
the properties such a hypothetical particle must have to fit
their bill. It (or they) has to be even more reluctant to
interact with baryonic matter than neutrinos are, interacting only very weakly indeed, except through gravity.
They must be stable, sticking around to dominate the Universe ever since the big bang. And, of course, the particles have to have some mass, in order to interact
through gravity at all. So they are often referred to as
"weakly interacting massive particles", or WIMPs, and
regarded as real candidates for the missing mass even
though they have never been detected. The terms WIMP
and CDM are synonymous and interchangeable, in this
connection.

Although WIMPs have never been detected, cosmologists seeking the missing mass were not entirely guilty
of pulling a rabbit out of a hat when they suggested the
Universe might be dominated by cold dark matter instead
of neutrinos. Particle physicists have been developing a
very successful unified theory of how the forces and particles of nature interact. This theory, called supersymmetry
(SUSY), accounts for what we know about the particle
world, but at the cost of requiring that for every kind of
particle we already know about there must be a
supersymmetric partner. For example, the electron has a
(hypothetical) counterpart dubbed the selectron; the
photon a partner called the photino; and so on. Most of
these particles are of no concern to the search for dark
matter, since they are unstable and will rapidly decay into
other particles (assuming they really do exist at all). But
the theory requires that there should be one type of SUSY
particle (the lightest member of the family) that *is* stable,
and stays around forever, and probably one that is very

long-lived. They would have very small masses, but it is a clear requirement of the basic theory that there *is* dark matter in the Universe, and that its overall contribution to the density cannot be ignored. There is also scope for a slightly different kind of particle, called the axion, whose presence would help the particle physicists to explain some subtleties of the interactions they observe, but which, like the SUSY particles, has never been directly detected.* Armed with all that information, many cosmologists have been happy to assume that WIMPs exist in sufficient numbers to close the Universe, even though no SUSY particle has yet been detected.

Assuming WIMPs do exist, what would a CDM dominated Universe look like? The models have no difficulty accounting for the existence of features like galaxies and clusters of galaxies. As the slow-moving WIMPs congregate in clumps in the expanding Universe, the clumps will exert a strong gravitational pull on any baryonic material nearby, acting like deep holes into which the baryons fall. The first clumps are smaller than in the hot dark matter scenario, so that galaxies are older than clusters, which are in turn older than superclusters. Which is where the model runs into difficulties. In the hot model, there doesn't seem to have been time for galaxies to form, but the large-scale pattern of bright matter across the Universe falls out fairly well from the equations. In the simple cold model, it is easy to form galaxies, but there hasn't been time for clusters of clusters of galaxies to group together in long chains and filaments. Perhaps the truth lies somewhere in between – although even the imaginations of the particle physicists haven't yet found a way to produce "warm" dark matter out of the big bang – or in a combination of two or more different kinds of dark matter. Or, perhaps, there is a different solution to the puzzle, a variation on the basic CDM theme.

When cosmologists map out the patterns of high and low density across the "Universe" in their computer models, they are really tracing the distribution of the dark

---

* For those with a taste for such things, a little more information about SUSY and the WIMPs can be found in the Appendix.

matter, which is 99 per cent of all the gravitationally important material around. Perhaps the dark matter is distributed more smoothly than the clumpy distribution of bright matter, with lots of unseen material in the voids between the chains and filaments of the froth. It may be that baryons falling in to gravitational pot holes only light up to form galaxies of stars when they fall into the very deepest pot holes, the regions where the dark matter is concentrated most strongly, and that in most of the Universe there are huge clouds of hydrogen gas gripped in the gravitational embrace of the WIMPs, but not strongly enough to trigger the processes that make galaxies.* It would help if we knew exactly what those processes were, but we do not. Even without that information, though, it is at least worth considering that by concentrating our attention on bright galaxies, we may be getting a completely biassed picture of the overall distribution of matter through space.

Until WIMPs are detected here on Earth, or (much harder) proved not to exist, or until completely new observational evidence comes in, many of these questions will remain open. But at the time of writing, late in 1986, it seems to me that the cold dark matter scenario, in one form or another, is by far the best model we have. It is certainly imperfect, and will be modified considerably in the years ahead. It is possible that it will turn out to be completely wrong, but that now seems very unlikely. And since space does not allow me to go into details of all the models of dark matter universes now being discussed by the theorists, it seems best to concentrate on the front runner, with the warning that the race is not yet over, that the front runner may fall, and that al-

---

* When astronomers study the light from very distant quasars, they find that the bright spectrum of the quasar is crossed by many dark lines corresponding to many "copies" of the spectrum of cold hydrogen, but shifted by different amounts to the red. This is called the Lyman forest; it is interpreted as being caused by light from the quasar being absorbed in many different clouds of cold hydrogen at different distances (different redshifts) between us and the quasar. These clouds could be "failed" galaxies, masses of hydrogen trapped by WIMP potholes but which never formed bright stars. If that interpretation is correct, the voids between the froth of bright galaxies may indeed be full of cold, dark matter.

though my personal prejudice is in favour of the WIMP model (for reasons that will become clear in Chapter Seven), I have been known to back astronomical losers in the past!

A BIAS FOR COLD, DARK MATTER

The way astronomers test their ideas about galaxy formation in models of the Universe dominated by different kinds of dark matter is with computer simulations. We can see the pattern of galaxies on the sky, and the kind of chains and filaments the pattern produces. In the computer models, equations are set up to describe points representing galaxies, themselves set up in a uniform three-dimensional cubic grid. The computer program then does the numerical equivalent of displacing the "galaxies" slightly from their starting positions and allowing them to interact in accordance with the law of gravity, while the grid is expanded to simulate the expansion of the Universe. The interactions between the "galaxies" can be altered by putting in the effects that would be produced by hot, cold or warm dark matter, and after a suitable interval of computer time, representing the evolution of the Universe up to the present day, the new positions of the galaxies, now distributed far from uniformly, can be shown on a screen, or printed out, as representations of how the sky would look in such a model Universe. At a superficial level, the patterns that are most like those of the real sky can be picked out by eye; more subtle tests, involving precise statistical comparisons of the nature of the chains and filaments that are produced by the computer simulations with the chains and filaments of the real world, give the ultimate test of whether each model is a good or bad approximation to reality.

All this takes up a great deal of computer time. There are several groups involved in such work; I discussed the

basic techniques with Carlos Frenck, a Mexican astronomer now at the University of Durham, in England, who
has carried out such studies in collaboration with George
Efstathiou of the University of Cambridge, Marc Davis of
the University of California, and Simon White of the University of Tucson. One of their recent studies involves a
grid of 32,768 "galaxies", perturbed by the appropriate
interactions. This may seem like a modest representation
of the real Universe, where over a million galaxies can
easily be pictured by putting together images from different
plates of the northern sky alone (see page 144). But it is
the best that can be achieved with modern computing
power, and it does give a powerful insight into how the
Universe works.

These "N-body" simulations (in this example, "N" is
32,768) mesh in neatly with the calculations made by
the theorists of how matter ought to behave in a Universe dominated by WIMPs. Objects with masses in the
range of a hundred million times the mass of the Sun
to a trillion (a million million) times the mass of the
Sun would have formed quickly in a CDM Universe, and
this is just the range of masses occupied by known
galaxies, which are indeed just about as old as the Universe
itself.

The galaxies are supposed to have formed from local
density fluctuations in the sea of WIMPs, producing gravitational potholes which trapped the baryonic material.
Most galaxies are spirals, like our own, and presumably
correspond to modest, but relatively common, fluctuations.
Giant, elliptical galaxies are rare, certainly no more than
15 per cent of the total, and would have formed from
larger, but rarer, random fluctuations in primordial density. This is not just a rule of thumb. There are precise
mathematical equations describing the nature of such
random fluctuations, and the exact proportions of large
and small galaxies turn out to follow those statistical rules
very well indeed.

The theory even predicts that as baryonic material falls
into a pothole and begins to form stars the gravitational
interaction between the baryonic material and the dark
matter of what is now a galactic halo will produce an

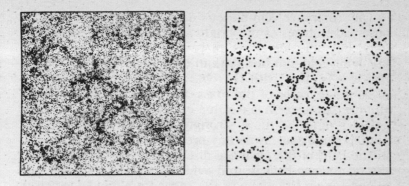

Figure 6.7 / When the computer simulations are carried
out using cold dark matter and with omega equal to
one, the picture begins to look very much like the real
Universe. The overall distribution of matter, including
the CDM, in such a Universe is shown on the left; the
distribution of "galaxies", represented as regions
where the overall density reaches peak levels, is shown
on the right. (These are the "N-body" simulations
described in the text; figures supplied by Carlos
Frenck.)

orbital velocity in the developing galaxy that is the same
right across its disc, exactly as observed.

Now the observers come into the picture. The cosmic
background radiation is very uniform, homogeneous to
within one part in ten thousand. It last interacted with
matter 100,000 years after the big bang itself, so we know
that matter was equally uniformly distributed at that time.
In that case, the voids *cannot* be empty, because even ten or
twenty billion years is insufficient time to sweep matter out
of them so drastically as to produce a Universe as in-
homogeneous as ours *seems* to be, judging by the distribution
of bright galaxies. Mass does not follow the light, and some-
how galaxies must have formed only where the density
fluctuations were at their biggest, leaving many failed galaxies scattered across the voids. Most matter is distributed in a
vast unseen ocean, much more evenly than the visible galaxies; even a huge supercluster of galaxies is only the equivalent of a small ripple on the sea of dark matter.

The computer simulations can reproduce all of the observed features of the bright galaxy distribution beautifully, but only for either of two possible scenarios, each involving cold, not hot, dark matter. In the first, the model Universe is open, with $\Omega = 0.2$ and the galaxies assumed to be a fair tracer of the mass. In the second, $\Omega = 1$, the Hubble parameter has a value of 50 km/sec/Mpc, and the galaxies must be more strongly clumped than the dark matter. Quite apart from any prejudices in favour of $\Omega = 1$ in the light of the evidence discussed earlier in this book, the microwave background evidence seems to tilt the balance strongly in favour of the second option. As icing on the cake, when Frenck and his colleagues looked at galaxies themselves they found more evidence that they were on the right track. Once the details of the model are fixed by making it match up with the observed Universe, these details can be used as parameters which limit the behaviour of galaxies within that model, in a new set of computer simulations. The simulation with $\Omega = 1$ and biassing automatically produces the massive halos and flat rotation curves that were such a surprise to astronomers just a few years ago, with no further effort by the modellers. Frenck stresses that this is "quite remarkable" since

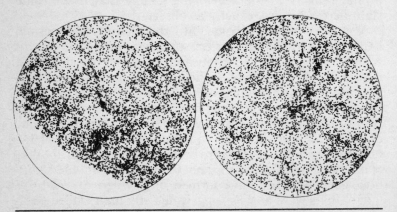

Figure 6.8/ A direct comparison of the real distribution of galaxies on the sky (left) with a computer simulation for $\Omega = 1$ in a cold, dark matter Universe. (Figures supplied by Carlos Frenck.)

the "free parameters of the model had been fixed beforehand by observations". All that's left, it seems, is to explain why galaxies formed in this biassed way, and failed to form in the voids.

Astronomers only realised the need for such biassing to be operating very recently, so very little work has been done on what the processes involved might be. But as soon as they began to devote a little thought to the problem, theorists realised, once again, that the observations were rubbing their noses in something that ought to have been obvious all along. When the first galaxies formed, in the biggest gravitational potholes, they may well have exploded into life as stars lit up one after another. This process of primordial star formation could have sent a blast of energy outwards from each young galaxy, energy carried both in the form of electromagnetic radiation (heat, light, ultraviolet radiation, X-rays and so on) and as a blast wave through the material of intergalactic space. Both ultraviolet radiation and a "wind" of energetic particles ejected from forming galaxies turn out to be quite efficient at suppressing the formation of other galaxies over a range of 20 Megaparsecs or more. The trick works by providing more energy for each of the baryons in the intergalactic medium – by making the material hotter – so that it does not settle into the potholes so easily. All it needs is for less than 10 per cent of the total baryon mass to be converted into galaxies early on, in the deepest potholes, with less than 1 per cent of the energy produced as a result spreading out to the baryons in the voids.

Martin Rees, of the University of Cambridge, has stressed that these requirements are easily met, and that the proposed mechanisms are not weird inventions of fevered theorists trying desperately to shore up a shaky model. It would be astonishing, he says, if none of the processes now being investigated were important – that is, if there were *no* large-scale environmental effects that influenced galaxy formation. In spite of the uncertainties at present, there is certainly no need for proponents of the idea that omega is equal to one to be embarrassed by the frothy distribution of bright galaxies across the Universe, says Rees.

This is one area of uncertainty where further work is likely to strengthen the position of theorists who argue in favour of a closed Universe. It is only fair to mention, though, that there is one piece of very new evidence which *is* causing them some embarrassment. This is the study reported in 1986, and mentioned in Chapter Five, which suggests that our Galaxy and *all* of the galaxies nearby, covering a flat region like a plate 100 Megaparsecs across, are moving together through space in a vast stream, with a velocity of at least 600 kilometres per second relative to the cosmic background radiation. Other studies hint at equally large streaming motions in other regions of the Universe, and, taken at face value, these motions are difficult to reconcile with the basic biassed CDM model, which is otherwise so successful. No such large-scale streaming, for example, falls naturally out of the computer simulations so far carried out by Frenck and his colleagues.

It is far too early yet to know whether these discoveries will be confirmed, or will turn out to be mistaken in some way; and it is also too early to tell whether, even if they are real, they will continue to trouble the CDM model, or whether a way round the difficulty will be discovered. I warned you that this would be a book about research at the frontiers, without all the *is* dotted and the *t*s crossed. But the hint of a difficulty for my favourite CDM model is, perhaps, an opportune reminder that I should mention some of the other ideas, far stranger than WIMPs, which are waiting in the wings to come forward and provide the missing mass if WIMPs do prove inadequate to the task – or, perhaps, to help the WIMP models out of their difficulties by combining with the cold dark matter to explain details like the large-scale streaming velocities.

These are wild ideas, lacking a secure foundation in observational or experimental fact, but buoyed up by theoretical hot air. Nevertheless, just maybe one, or more, will turn out to have a role to play in the real Universe.

STRINGS AND THINGS

The most respectable of the wild ideas concerns the possible existence of cosmic string. But unfortunately, that terminology has been used in three different, but possibly related, contexts, so it's best to unravel the three uses of the term before getting on to the cosmological nitty gritty.

Working from the smallest scale upwards, many theorists are excited today about a relatively new theory of the particle world. Most of us have a mental image of particles like quarks and leptons as little round balls. For years, the physicists have been telling us this is wrong, and that we should think of these fundamental entities as mathematical points, occupying no space at all in any direction but extending their spheres of influence over the range of the fields of force associated with them.* Now, though, theorists seeking to unify all of the forces of nature into one mathematical description have found that they can overcome many of the difficulties by working in a 10-dimensional spacetime in which the fundamental entities are not represented by mathematical points but by tiny objects which have a one-dimensional length, little loops extending over about 10^{-35} (a decimal point followed by 35 zeros and a 1) of a centimetre. The "extra" dimensions, over and above our familiar three of space and one of time, are wrapped up in these tiny tubes. With grand hyperbole, the theory that builds from this, the smallest scale yet considered by scientists, is called "superstrings".

Superstring theory includes within itself the possibility of another kind of string, existing on a cosmic scale. But these cosmic strings, which are especially interesting in the context of the debate about dark matter, are not exclusively a product of superstring theory. Many other variations on the theoretical theme of the behaviour of matter and energy at the time of the inflationary era of the birth of the Universe produce the same prediction –

---

* Quantum physics, indeed, says that even the location of the mathematical point cannot be precisely known.

that there ought to be almost infinitesimally thin threads of mass-energy stretching across the Universe, each one about 10^{-30} cm across but enormously massive, containing about 10^{20} kilograms (a hundred thousand trillion tonnes) of matter in every centimetre of string. Such strings arc under equally enormous tension, and if they are stretched then the energy used to stretch them is converted into mass, so that they become even more massive.

The way in which such exotic objects are (might be) produced depends on the way the original unity of the Universe was broken up at the earliest times. The Grand Unified Theories (GUTs) of physics suggest that under conditions of very high energy there is no distinction between the fundamental forces of nature. Electromagnetism and the forces that dominate inside atomic nuclei, the strong and weak forces, are today very different from one another, and all three are different from gravity. But, as we saw in Chapter Three, theorists argue that at the moment of creation itself all four forces were equally strong and governed by one set of mathematical rules. As the Universe cooled, the different forces separated out and showed four different faces to the world, but those four faces should still be describable in terms of one unified mathematical package. The fact that theorists have not yet found the right mathematical description for that unified package is regarded as a minor inconvenience; they have several good lines of attack on the problem, including superstring theory, and they are sure they will crack it one day. The important point, as far as we are concerned now, is that in many variations on the grand unified theme the splitting off of the four forces as the Universe cools involves a change from one state to another, a fundamental alteration in the physical properties of the universe, which is called a phase transition. I mentioned earlier (page 49) that this is likened by physicists to the phase change that occurs when water freezes into ice. Now, it is time to go in to a little more detail.

When a liquid crystallises, changing into a solid, it often does so imperfectly. Instead of one uniform lump of solid crystal, there may be separate regions, separate domains within the crystal, which are perfectly smooth and uniform within themselves, but in which the atoms and molecules are aligned differently from the pattern in the domain next door. So there are boundaries between domains, faults in the crystal, sometimes called discontinuities. Cosmic strings are faults in the fabric of spacetime, caused by imperfections in the phase transition that cooled the Universe out of its grand unified state. In a sense, these narrow but enormously long tubes enclose regions of spacetime in which the rules of grand unification still apply, and where all the forces of nature are one. They were produced, according to the grand unified theories, a mere 10^{-35} sec after the moment of creation itself.

Like WIMPs, cosmic strings should have been produced in vast numbers in the big bang. Unlike WIMPs, they whip through the Universe like highly stretched rubber bands, twanging at close to the speed of light. But most of them would not have survived to the present day. Such strings have no free ends, but always loop back on themselves to keep the string closed off from the rest of the Universe; the loops, however, can stretch across the entire visible Universe. But when two pieces of cosmic string cross (either two separate pieces or a tangled loop from one long piece of string), they can make new connections and shed smaller loops. The vibrating smaller loops also radiate energy away by gravitational radiation, making the loops shrink. Small loops are cut off from large loops which are cut off from huge loops which are . . . but you get the picture. If cosmic strings exist, then the Universe contains loops of material with various masses, just right to act as seeds for the growth of galaxies through their gravitational influence on the surrounding sea of WIMPs and baryons. Small wonder that astronomers seeking to account for the presence of the third kind of universal string, the long chains of galaxies that I have been careful to refer to so far as filaments, have seized with delight on the idea of cosmic string. Could it be that superstring begat cosmic string which begat strings of galaxies? Even with all the uncertainties in the chain of argument, the possibility is intriguing enough to follow up.

The properties of cosmic string certainly seem to be just what the astronomers wanted. It is rather strange stuff,

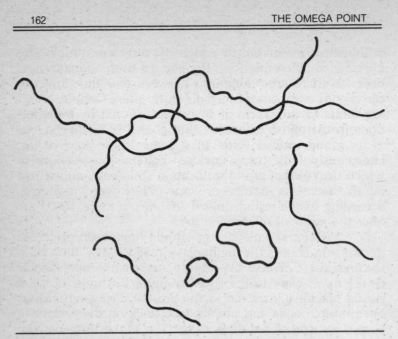

Figure 6.9 / Cosmic String
If two cosmic strings overlap, they can break and
reconnect to form loops. Could these be the seeds of
galaxies?

because in spite of its enormous mass a long, straight
cosmic string actually has only a small gravitational influence on its immediate surroundings today. In effect,
though, a straight string takes a slice out of spacetime as
it passes, altering the local geometry with bizarre implications for any object it happens to meet. Suppose a
string passed right through your body, horizontally at
waist height. You wouldn't feel a thing, at first. But in the
wake of the string, where spacetime is distorted, the top
and bottom of your body would be moving together at a
speed of about 4 km per second, with uncomfortable conse-
quences. Or imagine such a string passing close by,
through your room but not through yourself. Again, you
would feel no gravitational pull at all, and if your eyes
were shut you wouldn't know anything had happened –
until the wall on the opposite side of the room smacked
you in the face at a speed of several kilometres a second!

Strings moving through space leave a pronounced wake behind them, and it is very tempting to imagine that these wakes might account for the growth of galaxies in filaments and sheets across the Universe. Alexander Vilenkin, of Tufts University, said as much to a meeting debating the links between particle physics and cosmology, at Fermilab in Chicago in 1984: "A distinctive feature of the string scenario is the formation of planar wakes behind relativistically moving strings. These wakes can help to explain the formation of large-scale structure in a Universe dominated by weakly interacting cold particles, such as axions."

But there is a limit to how much cosmic string there can be around today. Anything that distorts spacetime bends light around itself, and distorts other forms of electromagnetic wave, such as the cosmic background radiation, passing by. Some astronomers hope to identify cosmic strings far away across the Universe by finding the influence – a kind of long, thin distorting lens effect – in the images of even more distant objects, such as quasars. They haven't yet succeeded, and perhaps there simply are not enough long pieces of cosmic string left for the effect to be noticeable. We know that the Universe is very uniform, from the microwave background evidence. That sets limits on the amount of cosmic string there is, since too much would produce bigger variations in the background from place to place on the sky than we actually see. Present observations of the background tell us that strings cannot contribute more than one hundred-thousandth of the total density needed to close the Universe. But even a fraction one-tenth smaller than this, with strings providing only one millionth of the closure density, would still be enough for them to leave their mark by producing galaxies in a Universe dominated by cold dark matter.

The details are still being worked out, and new ideas are aired almost every month. But it does seem that loops of string might help galaxies to grow. The smallest loops around today would weigh in at about the mass of a billion Suns, and encompass a diameter of about 10,000 light years; such small loops affect spacetime rather differently from a long, straight string, and would indeed tug in matter, both WIMPs and baryons, like water running into a lake. Even better, according to calculations made by Neil Turok, of Imperial College, in London, the way large loops split off smaller loops, and the smaller loops then cluster together, fits very neatly the pattern of characteristic clusters of galaxies, named "Abell clusters" after their discoverer, which typically contain 50 or more galaxies grouped in a volume of space only some 1.5 Megaparsecs (just less than five light years) across. Even before anyone has yet carried through the necessary calculations to establish the case, it is natural to speculate that bright galaxies may form only in frothy sheets and filaments across a Universe full of cold dark matter because those frothy sheets and filaments happen to be the places where cosmic strings have passed by, and where loops of string remain today at the hearts of galaxies and galaxy clusters.

Cosmic string can even revive the hot dark matter scenario. Although small-scale variations in the density of baryons are still wiped out by free-streaming neutrinos (or other HDM particles) even if the Universe contains cosmic strings, as soon as the neutrinos cool to the point where they are unable to maintain the smoothness both baryons and neutrinos can begin to accumulate around loops of string, forming galaxies much earlier than in an HDM model without string. Division of string loops could, perhaps, explain why some galaxies seem to be in pairs that have split apart, like amoebas; although nobody has yet carried the necessary calculations through, there is speculation that the disturbances produced in the wake of a passing string could account for the large-scale "streaming" velocities of galaxies, like those recently found in our vicinity; and the now obvious connection of many galaxies in chains and filaments – the third kind of string – could also be explained by the formation of a line of "seeds" as an original long, tangled string doubles over itself and splits off loops. It will be especially interesting to see how the N-body simulations come out once the effects of string are included in the computer programs; meanwhile, it is hardly surprising that astronomers are excited about cosmic string, which is certainly the current flavour of the month. A combination of strings *and* WIMPs (or even neutrinos with mass) reproduces so many features of the observed Universe that it is tempting for theorists to believe that they are on the right track at last. And they still have other cards up their sleeve.

BLACK HOLES, QUARK NUGGETS AND SHADOW MATTER

Apart from changing the rules of the game, by saying we don't understand gravity after all, or invoking the presence of something we don't know anything about (both desperate councils of despair), there are three more cards the theorists have ready to play if needed. One sounds familiar, but appears in an unfamiliar guise – the concept of black holes.

Black holes would certainly be good candidates for dark matter, and do indeed influence their surroundings through gravity, as required. But "ordinary" black holes, made when stars die and collapse in upon themselves, or when matter funnels on to a supermassive core at the heart of a galaxy or quasar, cannot be invoked to provide the missing mass, since they are originally composed of baryons themselves. Even if they are later crushed out of existence inside black holes, the baryons were still manufactured in the big bang, and are still subject to the limits set by the helium abundance. The only way that black holes could provide the dark matter needed to close the Universe, without running into this limit on baryon numbers, is if the black holes formed *before* the baryons were cooked, even earlier in the big bang. Such primordial black holes would be much smaller than atoms, and would each have about the mass of a planet. They *could* form from density fluctuations in the very early Universe, before the epoch of baryons, but the only reason for invoking their existence is to provide the missing mass. There is no observational evidence that they exist, and they do not naturally produce the required frothy distribution of galaxies in the real Universe. As a proposal for the dark matter that dominates the Universe, they scarcely rank above invoking magic, or mysterious unknown phenomena.

If you have a taste for scientific ideas that smack of magic though, the theorists can provide a much more entertaining possibility than black holes. One variation on the superstring theme, which includes gravity and all the other forces in one mathematical formalism, contains an extra kind of splitting over and above the symmetry breaking that is required to divide the original grand unified force into four components. According to this version of the equations, there was another splitting, even earlier in the big bang, which produced two separate sets of particles and forces, each occupying the same Universe. One set is our familiar world of stars, galaxies, planets and the rest, held together by the familiar four forces. The other set is – something else, invisible and perhaps undetectable, co-existing with us in our Universe but with its own particles and forces, interacting only with each other, not with our world.

What would this shadow world be like? It is possible, though unlikely, that it could be a kind of duplicate of our Universe, with the same, or similar, four forces, particles equivalent to quarks and leptons, and shadow stars, planets and even people going about their business and (perhaps) speculating about the possibility that *our* world exists. It is far more likely, however, that *if* the shadow world exists then the laws of physics will be slightly different, or very different, there. It could contain different particles that obey different rules. But either way there is only one way in which the two worlds could interfere with one another, and that is through gravity. Could the missing mass be in the form of a whole alternate world, occupying the same space as our own? Could some of the dark matter of the Galaxy's disc be in the form of shadow stars and planets? Or might our whole Universe be filled with exotic shadow particles that contribute to the overall density of the Universe but play no part in the particle interactions that involve neutrinos, WIMPs, cosmic strings and the rest of our world?

This is far from being the end of the imaginings of the theorists, and some other wild ideas are mentioned in the Appendix. But shadow matter represents in many ways the most extreme possibility yet conjured up. The notion should be served up with a large pinch of salt, and the better we can explain the observed distribution of galaxies without invoking shadow matter, or other exotic possibilities, the less likely it is that this bizarre vision will turn out to have any practical relevance. Even so, it is comforting to be reminded that if those other ideas, that look so strong today, do turn out to be flawed, there are still theorists around with imaginations vivid enough to dream up new possibilities.

Somewhere on the plausibility scale in between the wild idea of shadow matter and the relative respectability of WIMPs is another theoretical speculation, the strange matter that I mentioned in passing earlier. From one point of view, this is a rather conservative idea, since it does not require the presence of any completely new particles, but only the kind of matter we already know about in a new, denser form. The idea which some theorists have followed up is that a lump of matter made up of roughly equal numbers of up, down and strange quarks might be stable. Familiar baryonic matter, remember, contains only up and down quarks; the name given to this hypothetical form of matter containing strange quarks as well is, logically enough, "strange matter". Hypothetical lumps of strange matter are sometimes called quark nuggets; there is no certainty that they would be stable – that depends on details of the strong interaction and other properties of quarks that are hard to calculate accurately. But if they do exist, quark nuggets contain more or less conventional matter which has never been processed into baryons as such, and so is not subject to the helium limit. Speculations about strange matter range from the possibility that nuggets with mass less than twice the mass of a proton might have been left over from the big bang, and provide the missing mass, to the idea that whole stars of strange matter might exist today, produced by the collapse of old, dead stars through a neutron star state and on into a strange matter state. Even if strange matter exists, however, there is no natural way in which it could account for all of the dark matter required to close the Universe.

The best candidates for the missing mass are not the ones dreamed up by the theorists, but the ones forced upon us by observations and experiments. WIMPs are *required* by the N-body simulations, for example, in order to provide a match with the real world; and cold dark matter is also the best candidate to explain other puzzling astronomical observations, as I describe in the next chapter. The rest, interesting though it may be to fans of science fiction, is largely conjecture. So perhaps, having wandered so far off the beaten track, I should outline the current cosmological "best buy" before looking at that dramatic new evidence, from much closer to home.

THE BEST BUY

Our Universe contains matter, some of which we can see in the form of bright galaxies of stars, but most of which is dark and cannot be directly detected. Galaxies are grouped together in clumps, much bigger than astronomers had suspected until recently. An individual supercluster can extend across 500 million light years, and some filaments are so long that a few astronomers believe they may be just the visible portions of strings of galaxies that stretch across the entire Universe. Galaxies within superclusters move much more rapidly compared with one another than anyone had expected, and large numbers of galaxies move together with velocities of hundreds of kilometres per second relative to the cosmic background radiation, quite separately from their motion due to the expansion of the Universe.

The "best buy" models of the Universe can explain all this. We do not know for sure what 99 per cent of the cosmic mass consists of, and it is likely that it is not all the same stuff. "Ordinary", baryonic dark matter, for instance, is quite sufficient to explain the way stars concentrate in the plane of our own Galaxy, with brown dwarfs and Jupiters doing their gravitational bit to keep things held in place. But when the growth of structure in the Universe is traced by N-body calculations in the biggest computers available, it turns out that while neutrinos might just be made to fit the bill (perhaps with the aid of cosmic string), most of the observed features of the galaxy distribution fall naturally into place if the Universe is just closed (with omega equal to one) and the hidden mass is in the form of weakly interacting cold dark matter. The galaxy formation process has to be "biassed" to account for the observed frothy appearance, but there are several likely mechanisms to produce biassing, even without invoking cosmic strings. Although cosmic strings are not strictly needed, they do fit neatly into the CDM scenario, and there are hopes that they may explain the large-scale streaming motions and the tendency of galaxies to congregate in sheets and filaments.

The most dramatic feature of these new discoveries is the finding that the smallest objects in the Universe, elementary particles, are responsible for the structure we see in the largest objects in the Universe, superclusters of galaxies. As things stand today, progress by astronomers towards a better understanding of galaxies and clusters of galaxies is impeded not by any lack of astronomical observations and information, but by the limitations imposed by our uncertain understanding of the basic physics involved at the particle level. This situation almost exactly echoes the problems facing physicists more than half a century ago, when astronomical observations and astrophysical calculations could tell them exactly how hot the Sun must be in its heart, but knowledge of nuclear physics was inadequate to explain the reactions going on inside the Sun which could maintain that heat for thousands of millions of years. And that is more than just a pleasing echo of the past, since by one of the greatest of astronomical ironies it turns out that astronomers seeking to determine the nature of the missing mass by studying the distribution of galaxies in the most remote depths of space might

have found what they were seeking more quickly if they
had concentrated their attention in our own astronomical backyard, on the Sun itself.

CHAPTER SEVEN

THE SOLAR CONNECTION

During the 1920s, astronomy and physics began to come together in a new scientific discipline, astrophysics. Spectroscopy told astronomers what stars, including our Sun, were made of; measurements of the orbits of binary stars revealed how much mass some stars contain, and techniques for measuring the distances to some of those same binary stars enabled astronomers to work out the true brightness of these objects. Brightness and mass turned out to be related, in a straightforward way, and this gave insight into the physics of stellar interiors. The mass of our own Sun is known very accurately, from studies of the orbital dynamics of the Solar System and the strength of tides raised by the Sun on Earth; its distance is measured by a variety of geometrical techniques; and its brightness can easily be determined from observations. Putting all the evidence together, pioneering astrophysicists, led by Arthur Eddington in Cambridge, found that the relation between the Sun's mass and its luminosity exactly fitted the rules that seemed to apply to other stars, and they calculated how hot the Sun must be at its heart, in order for the pressure inside it to be strong enough to hold the outer layers up against the inward tug of gravity. This is simple physics, applied to stars – astrophysics. It told Eddington and his colleagues, just as it tells astronomers today, that the temperature at the centre of the Sun must be close to 15 million degrees, Celsius. And while studies of the Sun as it is today were beginning to yield up precise information about its interior, geologists studying the rocks of the Earth were pushing back the known age of our planet, and of the Solar System, deducing that the Sun must have been shining in much the same way that it does today for 4,000 million years or more, in order to provide the time needed for geological processes to shape the Earth.

THE HEAT OF THE SUN

There is only one possible source of energy that can keep the centre of the Sun so hot, and that has been able to keep it hot for the thousands of millions of years required by geology. This is the conversion of matter into energy, in line with the familiar equation from special relativity, $E = mc^2$. To be precise, in order to maintain the Sun's present luminosity, pouring energy out into space at a rate of 4×10^{33} ergs per second, just over four million tonnes of matter must be converted into pure energy every second. This is only a tiny fraction of the Sun's mass, which is 2×10^{27} tonnes. Provided there is a process for turning the mass into energy, there is no shortage of "fuel" to keep the Sun hot for the the thousands of millions of years required. But what is the process that turns matter into energy inside the Sun?

In the late 1920s, the astrophysicists thought they knew. Under the extreme conditions of heat and pressure inside the Sun, there are no hydrogen and helium atoms as such. The electrons are stripped off from the atomic nuclei, which are left bare, and nuclei and electrons intermingle in a hot fluid called a plasma. "Fluid" is perhaps a misleading description, though, since this is like no fluid we ever experience on Earth. The density of the plasma at the heart of the Sun is twelve times the density of lead, so the atomic nuclei have plenty of opportunity to interact with one another, each one being bombarded by the others in a constant high temperature, high pressure battering. Most of these nuclei are simply protons, hydrogen atoms stripped of their single electrons. About a quarter of the baryons are in helium nuclei, left over from the big bang. Some are in helium nuclei that, astrophysicists believe, have been made inside the Sun itself. And there are traces of heavier elements, formed in earlier generations of stars and spread through space, part of the cloud of material from which our Solar System formed.

A helium-4 nucleus consists of two protons and two neutrons. Since a neutron left in isolation will decay into a proton and an electron, it seemed a reasonable guess that under the conditions at the heart of the Sun four protons and two electrons from the hot, dense plasma might, from time to time, be squeezed together to make one nucleus of helium-4. This would be just what the doctor ordered, since the mass of a nucleus of helium-4 is less than the mass of four protons and two electrons added together. If 600 million tonnes (6×10^8 t) of protons (hydrogen nuclei) were converted into helium-4 nuclei inside the Sun every second, the mass converted into energy in the process would be just the 4 million tonnes or so required – about 7 per cent of the "fuel" is turned into pure energy by the fusion process.

The figures hung together so well that the astrophysicists were convinced. But there were, initially, difficulties with the explanation of how the Sun stays hot. Helium-4 nuclei are indeed stable once they are made, and it sounds easy enough to talk of squeezing four protons together, and adding in a couple of electrons (or emitting two positrons, the antimatter counterparts to electrons), to make helium nuclei. In practice, though, there is a problem with the protons' positive charge. Like charges repel, and even under the extreme conditions at the heart of the Sun the positive charge the protons carry makes it difficult for them to get close enough to each other to come into direct contact. Instead, the fast-moving, hot protons bounce apart without touching, in much the same way that two bar magnets, pushed together with north poles facing each other, will try to spring apart. This barrier between the protons can be overcome if the protons are moving fast enough – that is, if the plasma is hot enough. But when particle physicists first calculated how protons would interact under the conditions the astrophysicists said must exist at the heart of the Sun, they decided that fusion could not occur, because the temperature was too low. The fact that the Sun and stars do exist, and must be getting their energy from somewhere, seemed to fly in the face of basic particle physics; but it turned out that the astrophysicists were right all along, and the particle physicists were the ones that had to rewrite their theories.

As quantum physics developed in the late 1920s and early 1930s, physicists realised that because of quantum uncertainty effects protons could indeed interact inside the Sun at a temperature of "only" 15 million degrees. The uncertainty in the positions of two protons that come close to one another can have the effect of smearing out the range of distance over which they can interact, in a sense making the two protons overlap with one another when classical physics would describe them as not even touching. The modifications to the classical picture required by quantum physics turned out to be exactly what the astrophysicists required, and this was one of the first practical demonstrations of the power of the quantum theory in describing the real world.

Of course, there were still complications, and a long road to travel before, in the 1950s, astrophysicists could at last say that they really understood how hydrogen is converted into helium inside the Sun. It isn't as simple as four protons coming together simultaneously – a rather unlikely event, after all – but involves the build-up of helium nuclei step by step. The process now thought to dominate inside the Sun, producing 98.5 per cent of its energy, sees first two protons combining to make a deuterium nucleus, spitting out a positron in the process, then the deuterium capturing another proton to make a nucleus of helium-3. Two

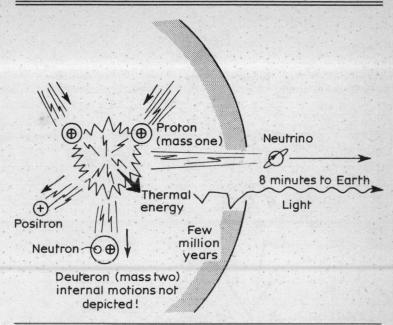

Figure 7.1 / Making Solar Neutrinos
When two protons combine to make a deuteron, one of
the products released is a neutrino.

helium-3 nuclei can then combine to make one helium-4
nucleus and spit out two protons back into the plasma.
Alternatively, one helium-3 nucleus may combine with
one of helium-4 to make a nucleus of beryllium-7. This
may lose an electron to become lithium-7, then gain a
proton and split into two helium-4 nuclei, or the beryllium-
7 may absorb a proton to become boron-8, which spits out
a positron to become beryllium-8 and *then* splits into two
nuclei of helium-4. Whichever route is followed, the net
effect is the conversion of four protons into one nucleus of
helium-4. Don't worry about the details, but remember the
boron and beryllium – they turn out to be particularly interesting.

   Some of these steps in the chain also involve the production of neutrinos. Every time a positron or an electron
is involved, in fact, there is a neutrino (or anti-neutrino)
produced as well, keeping the total number of leptons in

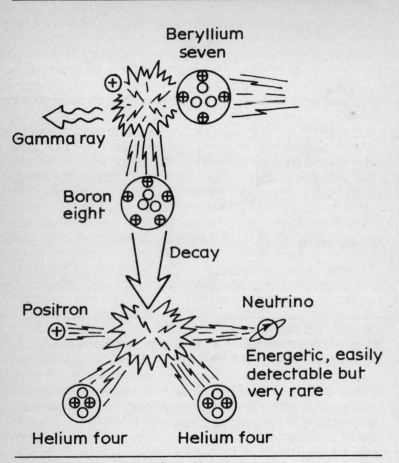

Figure 7.2/ Making More Solar Neutrinos
Another process going on inside the Sun which,
according to theory, ought to be releasing neutrinos.
Beryllium and boron nuclei interact to make two
helium-4 nuclei and release one neutrino. Why do we
see only one-third of the expected number of these
high-energy solar neutrinos?

(Figures 7.1 and 7.2 based on diagrams supplied by
Willy Fowler)

balance. In shorthand terms, the net effect of the fusion
process is to convert four protons into one nucleus of
helium-4 *plus two positrons and two neutrinos.* When two

protons combine to make deuterium, they release a neutrino along with the positron they spit out; when beryllium-7 absorbs an electron (which is equivalent to emitting a positron), it spits out a neutrino; and every time a nucleus of boron-8 spits out a positron as it converts itself into beryllium-8 (which in turn splits into two nuclei of helium-4), it ejects a neutrino along with the positron. But neutrinos, as we have seen, are very reluctant to interact with anything. Neutrinos manufactured in these reactions at the heart of the Sun would stream out through the Sun itself, scarcely noticing the existence even of matter with 12 times the density of lead, and would pour across space, past and through the Earth.

In the 1960s, astronomers realised that if they could only find a way to detect these solar neutrinos, and measure their flow, they would have, in effect, a direct window opening on to the centre of the Sun, a means of monitoring the processes that keep stars hot inside. The very reluctance of neutrinos to interact with ordinary matter, however, made the prospect of designing and building a detector to trap solar neutrinos a daunting one. Since "everyone knew" what kept the Sun hot, there was no pressing incentive to carry out the task. But one man was not daunted by the difficulties, and did carry through a project to build a solar neutrino detector, in the face of much wry amusement from colleagues who felt that since either the detector wouldn't work, or it would only tell us what we knew already, it wasn't really worth the effort. During the 1970s and into the 1980s, however, the detector *was* made to work, and did *not* show the expected flow of solar neutrinos. Instead, during what is now a long series of very reliable observations, it has found just one-third of the predicted number of solar neutrinos. Either we do not, after all, understand the astrophysics, or there is something wrong with conventional particle physics. The echo of the 1920s rings loud and clear, with implications that are, this time, of literally cosmic importance. Before we look at the implications, and the possibility that the neutrino detector may actually have found the best evidence yet of the nature of the missing mass, we need to understand exactly what it is that a tank full of cleaning fluid, buried down a gold mine in South Dakota, is telling us about the Sun.

THE SOLAR NEUTRINO PROBLEM

Since the mid-1960s, Ray Davis, of the Brookhaven National Laboratory, has devoted his scientific career to the search for solar neutrinos. Because neutrinos are so reluctant to interact with anything, he wanted to use in his detector a material that was cheap enough to be affordable in large quantities. He chose chlorine, in the form of perchloroethylene (carbon tetrachloride, C_2Cl_4) which is widely used for cleaning clothes (so-called "dry cleaning") and is easily available from industrial suppliers. Chlorine, like other elements, comes in different varieties, or isotopes, and on average one chlorine atom in four is the isotope chlorine-37, which can absorb a neutrino to create an atom of argon-37. On average, one of the four chlorine atoms in each molecule of perchloroethylene will be chlorine-37, and able to take part in this reaction. If the reaction does take place, the jar produced by the impact of the neutrino will knock the argon-37 atom out of the molecule, and in principle the argon-37 produced in a tank of cleaning fluid in this way can be identified by chemical means. But – you need a *lot* of cleaning fluid to have any hope of capturing neutrinos from the Sun.

If the standard models of how the Sun works are correct, the proton-proton (p-p) interactions should be producing a flood of 60,000 million neutrinos crossing every square centimetre of space at the orbit of the Earth in every second. Unfortunately, these neutrinos are produced with relatively low energy, and are not potent enough to trigger the chlorine-37 reaction. The beryllium-7 neutrinos, produced in one of the possible routes to helium-4, have more energy, and ought just to trigger the chlorine/argon switch. But since only a small fraction of the helium is produced in this way, the theory says that there are only 4,000 million of these neutrinos crossing every square centimetre of the Earth every second. Boron-8 neutrinos are scarcer still, just 3 million per square centimetre per second, but even more energetic and therefore easier to detect. It was these higher energy neutrinos that Davis expected to find. But first he had to take precautions against other particles from space, cosmic rays, triggering the switch from chlorine to argon.

In order to shield his detector, Davis had it built 1,500 metres below the surface of the Earth, in the Homestake Gold Mine at Lead, South Dakota. With all that rock above to block out anything except neutrinos, he still needed 400,000 litres of cleaning fluid, in a tank the size of an Olympic swimming pool, to have any realistic hope of detecting anything at all. The theorists calculated that, after all that effort, the tank would capture just 25 to 30 solar neutrinos each month. In a different notation, the physicists involved in this work talk in terms of "solar neutrino units", or SNU (pronounced "Snew"). These are defined in terms of the flux of neutrinos and the probability that one neutrino will be captured by one of the particular type of "target" atoms in the tank; the details are not important, but what matters is that the standard models of astrophysics and particle physics imply that Davis's tank should register 6 SNU. In a series of more than 50 test runs, starting in 1968 and continuing up to date, he has consistently found a capture rate of about nine neutrinos per month, corresponding to 2 SNU. After exhaustive tests on the systems used to detect the argon atoms produced in the tank, and checking out every other possibility, the physicists are left in no doubt that the detector really is recording only one-third of the expected number of neutrinos from the Sun. This is "the solar neutrino problem".

Taken at face value, the Davis experiment says that either our theory of stellar astrophysics is wrong, or we don't understand the way neutrinos are produced in particle interactions (or both!). This has encouraged theorists to come up with many speculative ideas about how to suppress the production of solar neutrinos, by changing

the astrophysics, or the particle physics, or both. John
Bahcall, of the Institute for Advanced Study, in Princeton,
says that between 1969 and 1977 he and Davis counted
nineteen independent suggestions of this kind, then gave
up counting, while "new ideas have been suggested at the
ratc of two or three per year since then". The vital need,
to cut the rug from under most of the speculations and
give a clue to what is really going on inside the Sun, is a
new experiment. After all, the chlorine detector mostly
captures neutrinos from the rare boron-8 process, which
is involved in the manufacture of only one ten-thousandth
of all the helium being made each second inside the Sun
today. A detector that could monitor the flux of the vastly
more prolific low-energy neutrinos produced by the p-p
process is bound to shed new light on this strange discovery. Almost all the astronomers, physicists and chemists
who have looked at the problem agree that the ideal
"second generation" solar neutrino detector will involve
gallium – 30 tonnes to build a really good detector, 15
tonnes to make some progress. It has all the required properties, both for absorbing the low-energy solar neutrinos
(at an expected rate of 120 SNU) and in producing an
easily monitored chemical change when a neutrino does
register in the target. Gallium, however, is *very* expensive
– hundreds of thousands of dollars per tonne – and is available only in small quantities – a few tonnes per year,
worldwide. Until the ideal experiment can be financed, the
astronomers and the rest will have to make do with other,
less than ideal, detectors, which are likely to become
operational within the next few years, but which may not
resolve the problem once and for all.

Davis has stirred up a hornets' nest with his discovery,
but as of now we cannot say for sure whether the resolution lies in a better understanding of the neutrinos
themselves, in changing the particle physics, or in
changing the astrophysics. Without going into details of
all the 50-odd speculations now aired to "resolve" the solar
neutrino problem, I'll attempt to give you a taste of each
of the three main lines of attack before spelling out the
details of the solution which so neatly fits the needs of the
cosmologists.

OSCILLATING NEUTRINOS?

Neutrinos, we know, come in three varieties, or flavours. There *might* be room in the Universe for a fourth flavour, associated with a fourth electron-like particle, but this seems unlikely. As things stand, the lepton family consists of the electron and its neutrino, the muon and its neutrino, and the tau particle and its neutrino. The neutrinos being produced by the hydrogen fusion reactions inside the Sun are all electron neutrinos (or electron *anti*-neutrinos, but the story is complicated enough without worrying about that). Davis's detector detects only electron neutrinos. Now, here is scope for a tempting speculation. Suppose something happens to the electron neutrinos *en route* to us from the Sun. Suppose, in particular, that they are transformed in some way, so that the original electron neutrinos are shared out evenly among the three possible varieties. Why, then we would see exactly one-third of the predicted number of electron neutrinos arriving at the Earth, just as Davis finds!

This is not quite the wild-eyed speculation it seems at first sight, because the rules of the particle physics game do allow certain kinds of particle to change their spots in this way. But there is one essential requirement before such spot-changing (technically known as an oscillation) can occur. The particles involved must have some mass. It doesn't matter how small the mass is, but it cannot be precisely zero. In the early 1980s, many physicists became quite excited by this realisation. A small mass for the neutrino could, perhaps, resolve the solar neutrino problem, by allowing oscillations. A small mass for the neutrino might, with a hundred or more neutrinos filling every cubic centimetre of the Universe, provide the cosmological missing mass. And when a handful of experiments began to come up with results consistent with neutrino mass and neutrino oscillations, joy was unconfined – but only briefly.

On the experimental side, Frederick Reines, Henry Sobel

and Elaine Pasierb, from the University of California at Irvine, monitored the behaviour of neutrinos produced in the Du Pont Company's 2,000-megawatt nuclear reactor at Savannah River. They set up a target, a pool of heavy water (deuterium oxide) just over 11 metres from the reactor, and in similar style to the Davis experiment counted the number of electron neutrinos arriving at the detector by monitoring the reactions produced in the heavy water. The experiment seemed to show that fewer electron neutrinos were reaching the detector than left the reactor – they were disappearing *en route*, just as, some suggested, solar neutrinos were disappearing. At about the same time, a team of researchers at the Institute of Theoretical and Experimental Physics in Moscow announced that they had found evidence of a mass of about 20 to 40 electron Volts for the neutrino (the mass of an electron is 511,000 eV).

Since then, however, the picture has become cloudy once again. Although other Soviet experiments come up with the same figure for neutrino masses, no western experiment has yet been able to find such evidence. A team at the University of Zurich, for example, recently came up with an "upper limit" of 18 eV, meaning that their experiment shows no signs of neutrinos with mass, and that it would, they believe, have detected anything above this limit. Meanwhile, cosmologists are worried that, with an electron neutrino mass as big as the Soviet results imply, and the masses of the other two flavours of neutrino assumed to be even bigger, there would be *too much* dark matter, making the Universe "over closed" and subject to a strong gravitational deceleration that we do not in fact see. And similar experiments to the one carried out at Savannah River have failed to confirm the existence of neutrino oscillations over short distances.

But this is not quite the end of the neutrino oscillation story. First, some of the most favoured of present theories, the GUTs, *require* that neutrinos have mass, however small that mass may be. And the sensitivity of tests like the Savannah River experiment depends on the ratio of the energy of the neutrinos to the distance between the source and the detector. For a given energy, the effects of a very small difference between the masses of the three flavours of neutrino will only show up over a very long distance – such as, perhaps, the distance from the Sun to the Earth. And some of the theorists who reject the idea that massive neutrinos could dominate the observed clustering of matter in the Universe, and who invoke axions, for example, to explain how galaxies cluster, would be quite happy with a background of rather light neutrinos, with a mass of one or two eV, perhaps even less, to help out on the still larger scale. The idea of neutrinos changing their spots between Sun and Earth has lost its original appealing simplicity, but cannot yet be ruled out. In 1986, though, theorists became excited about another variation on the theme, finding a way to allow the spot-changing to take place *inside* the Sun, before the neutrinos get out into space. The idea is a high-flyer at the time of writing, but might soon be shot down. Even so, it will serve as an example of the kind of approach to the solar neutrino problem advocated by those who seek to resolve the problem by changing the physics.

CHANGE THE PHYSICS?

The idea that neutrinos from the Sun might be changing their spots *inside* the Sun, instead of on the journey across space to Earth, first emerged in a scientific paper presented to an international gathering of physicists in Finland in 1985. The two authors, S. P. Mikheyev and A. Yu. Smirnov, hail from the Institute for Nuclear Research in Moscow; they jumped off from some earlier calculations concerning neutrino oscillations, made by Lincoln Wolfenstein, of the Carnegie-Mellon University in Pittsburgh, and the new idea is sometimes referred to as the "MSW" theory. But nobody got excited about it until Hans Bethe, one of the pioneers of the theory of nuclear fusion in stars, put the weight of his name behind it in a paper published in the journal *Physical Review Letters* early in 1986.

Bethe's endorsement made people sit up and take notice because, starting in the late 1930s, he had been the leading light in the development of an understanding of the details of the fusion processes which keep stars hot. His first idea, involving carbon nuclei as catalysts, turned out not to be very important for the Sun, although it does dominate the production of energy in more massive, hotter stars; with a colleague, Charles Critchfield, Bethe later came up with the p-p chain, which *is* thought to be the dominant energy source of the Sun.

The neat way in which the wheel had turned full circle, and the unusual sight of an eminent scientist coming back, in the year of his eightieth birthday, to tidy up details of a theory he had pioneered almost half a century before, ensured a wave of publicity in the scientific magazines for Bethe's statements that the Mikheyev and Smirnov contribution was "a very important paper" and that it "is the first explanation [of the solar neutrino problem] that could be right". Perhaps *too* strong an endorsement – I don't think the issue is yet quite that clear-cut. But certainly Bethe is telling us to take these ideas seriously.

The key feature of the MSW model derives from the description of neutrinos as waves, rather than particles. This duality is a fundamental feature of the particle world, and many experiments have shown that light behaves as a wave in some circumstances and as a stream of particles in other circumstances, while electrons, for example, sometimes interact in the manner of particles and sometimes in the manner of waves. The mathematical description of this strange behaviour is incorporated in the theory of quantum mechanics, a very good, successful description of the world of the very small. It is simply a fact of life, bizarre though it seems to our common sense, which is based on experience on a much larger scale, that an entity like a neutrino incorporates both wave-like and particle-like characteristics.* In these terms, the oscillation of a

---

* If you need convincing that this is indeed a good description of reality, check out my book *In Search of Schrödinger's Cat.*

neutrino from one flavour (one quantum state) to another can be thought of as a continuous tuning process, like tuning the dial of a radio continuously back and forth between three different stations, in which the wave shifts from the electron neutrino state to the muon neutrino state, to the tau neutrino state, and back again. So far, this is just another perspective on the old idea of neutrino oscillations. But now comes the twist.

Mikheyev and Smirnov pointed out that the oscillations will be affected by the way the travelling neutrino waves interact with matter on their way out through the Sun. To be sure, the interactions are very rare, but there is indeed a small chance that the part of the wave that corresponds to electron neutrinos will be affected by the matter the wave passes through – "scattered", in the terminology used by physicists. This happens for electron neutrinos, because there are plenty of electrons in the heart of the Sun to take part in the scatterings. But since there are no muons or tau particles there, there is no equivalent scattering effect for the part of the wave that corresponds to muon or tau neutrinos.

The effect of this scattering process is to increase temporarily the effective mass of the electron neutrinos involved, by giving them more energy, and to boost the chances of an electron neutrino "oscillating" into a muon neutrino, or a tau neutrino. This boost happens because the other two neutrinos are more massive than the electron neutrino under normal conditions, but the oscillation is strongest when the flavours involved in the oscillation have equal mass-energy. So the balance of flavours within the wave tilts away from electron neutrinos as the electron neutrino component of the wave gains mass-energy by scattering inside the Sun. Once the oscillation has taken place, however, the muon or tau neutrino components of the wave produced are unaffected by the matter it is passing through, and by the time the stream of neutrinos emerges from the Sun and the electron neutrino component ceases to be affected in this way the number of electron neutrinos has been depleted.

Could this depletion be sufficient to account for the shortfall of electron neutrinos detected by Davis? Bethe –

and several other researchers – calculate that the trick works in the right way if the difference between the quantum mass-energy states of the neutrino flavours is very small, less than about 0.008 eV, and if the probability of mixing occurring inside the Sun is less than 1 per cent. These numbers, which would imply that the mass of the muon neutrino is indeed about 0.008 eV and the mass of the electron neutrino is very much smaller still, are little comfort for cosmologists seeking the missing mass; to make omega equal to one with neutrinos alone, each neutrino must have a mass of a few tens of electron Volts. But the numbers are very much in line with the requirements that come out of the Grand Unified Theories of physics, which are among the most successful attempts to unify the forces and particles of nature in one mathematical description. So the particle physicists are, if anything, more excited by the MSW model than the astronomers are. The way to test the idea properly, however, is to build more solar neutrino detectors, designed to respond to neutrinos with different energies. Davis's detector, in effect, gives us just one point on a graph; with detectors operating at other energies, physicists will obtain a solar neutrino spectrum, and the way the neutrinos are distributed at different energies will very quickly tell us whether or not Bethe's enthusiasm for the MSW process is justified. If it is not, then it is going to be hard to avoid the conclusion that there is nothing wrong with our understanding of particle physics and neutrinos, but there is something seriously wrong with the standard theory of the Sun. And that, it turns out, really *would* help the search for dark matter.

CHANGE THE ASTRONOMY?

The simplest way to reduce the flux of solar neutrinos in line with the results of Davis's experiment is to turn down the temperature at the heart of the Sun by 10 per cent. It is a measure of how successful the theory of stellar structure has been that this modest proposal cannot be accommodated by the standard models, which say very precisely how much energy must be generated in the core of the Sun in order to give it its outward appearance – size, mass and luminosity. Whereas the particle physicists don't know if the neutrino has zero mass, or 30 eV, or something in between, astrophysicists are quite sure that there is no error as large as 10 per cent in their standard calculation of the warmth of the Sun. But they have had some fun speculating on how a star like the Sun might go off the boil, temporarily or permanently, if its regular pattern of behaviour were disturbed.

The point the speculators have seized on is that the Sun takes a long time to adjust to any changes that go on inside it. If the nuclear fusion reactions that keep it hot were turned off instantly, by magic, but the other laws of physics were unchanged, the Sun would not go out like a light being switched off. Energy produced in the heart of the Sun, in the form of photons, takes a very long time – in round terms, about a million years – to get out to the surface. Even though the photons are travelling at the speed of light, they cannot travel straight from the core of the Sun to the surface, but follow a zig-zag path of enormous complexity, bouncing from one atomic nucleus or electron to the next in the superdense core like a runaway ball in some cosmic pinball machine. Adding up all the zig-zags, a typical photon has actually travelled a million light years before it gets from the inside of the Sun to the surface, a straight-line distance of only 1,390,000 km; it then takes just over 500 seconds to cross the further 150 million km to the Earth.

So the light and heat by which we see and feel the Sun today is really a million years old. And even if the nuclear reactions had stopped a million years ago, we still would not see the Sun go out tomorrow. If a huge ball of hot fluid like the Sun gets cold, it will begin to contract. As it contracts, however, gravitational energy is turned into heat, and it warms up again. This is how a collapsing cloud of gas got hot enough in the middle to start up the nuclear reactions in the first place – and, the astrophysicists calculate, the Sun could maintain its present appearance for roughly 10 million years without burning any nuclear fuel at all. Only after all that time would we actually be able to tell that something had gone wrong, merely by looking at the Sun from the outside.

Even if all is well with the nuclear physics, the present appearance of the Sun is really a result of an average of all the nuclear reactions, and the processes of radiation and convection going on inside the Sun, over the past million years or more. The astrophysics says that *on average* the temperature at the heart of the Sun is 15 million degrees. If the temperature were to fluctuate by 10 per cent or so for a few years, or centuries, or even hundreds of thousands of years, we could never tell by looking at the Sun from outside. But the neutrinos, of course, fly straight from the centre of the Sun to Ray Davis's detector buried in the mine in South Dakota. Could it be that his observations were telling us that the Sun had gone "off the boil", temporarily, but the outer regions had not yet had time to adjust?

The idea gained some attention in the early 1970s, because it raised echoes of an earlier idea, introduced in the 1950s to attempt to explain the recurrence of Ice Ages on Earth. Ernst Öpik, an Estonian-born astronomer who was then working in Northern Ireland, suggested that during the normal phase of nuclear burning in a star like the Sun the heavier elements produced by the fusion reactions might build up around the hot core like a kind of nuclear ash. Eventually, he argued, this would become an even more efficient barrier than the usual core material is to radiation trying to get out from the core, and heat would be trapped at the heart of the Sun. This would make the outer layers cool and shrink slightly, while the core itself got hotter until strong convection currents were set up and burst out, carrying the "ash" outwards with them. While the star was recovering from this hiccup, the *core* would cool and nuclear reactions might slow down, or stop altogether, for a time.

It isn't easy to make this pattern of behaviour fit the known rhythms of ice ages, which have in any case now been explained rather satisfactorily without invoking

changes in the Sun. But the possibility of finding a way to
turn off the nuclear reactions for a while was the straw at
which astrophysicists clutched in the early days of their
concern about the Davis experiment. A rather nice variation on the theme speculated that the disturbing influence
might come from outside, rather than from within the Sun
itself.

From time to time, as it orbits around the Milky Way,
our Solar System passes through clouds of gas and dust in
space. As the Sun swept up material from such a cloud,
the argument ran, gravitational energy of the infalling
material would be turned into heat, just as it is when a
star shrinks a little, and this extra source of heat would
certainly disturb the convective patterns of the outer layers
of the star. Could the effects be felt in the deep interior,
and alter the nuclear burning processes? The possibility
was never more than pure speculation, but one which
gained a little credibility from suggestions that the Solar
System has recently (by astronomical time-scales) passed
through a region of space where there are such clouds,
and so the Sun might today be recovering from the effects
of such an encounter.

Other theorists threw their hats in the ring with suggestions that there might be natural fluctuations in the rate
at which energy is produced inside the Sun, with the
temperature at the core fluctuating as a result a little either
side of the long-term average of 15 million degrees. But
these were all seen as rather contrived "solutions" to the
puzzle, contrived to answer just the one problem of why
Davis should be finding so few solar neutrinos. There was
no natural explanation of the solar neutrino problem, no
model which also fitted in with other mainstream ideas in
physics, and which offered any predictions of other properties of the Sun and fundamental particles that could be
tested by new observations and experiments. None, that is,
until the triumphant arrival of the WIMP on the solar
stage.

ENTER THE WIMP

The solar neutrino problem may, in fact, have been solved as long ago as 1978 – but the two theorists who did the work did not publish their idea then, because they thought it was too bizarre to be taken seriously. The line of attack which John Faulkner and his student Ron Gilliland took up at the Santa Cruz campus of the University of California was the standard astrophysical approach of finding a way to make the core of the Sun 10 per cent cooler. But they tackled the problem not just by tweaking the astrophysics, but by invoking the presence of a hypothetical type of particle to do the job of carrying heat outwards from the centre of the Sun, and spreading the warmth through a small fraction of the Sun's inner radius. The neutrinos detected by Davis come from the innermost 5 per cent of the Sun's radius, where the temperature is highest. But most of the energy that keeps the Sun hot is generated over a larger volume than this. You can generate just as much energy overall as in the standard models by having

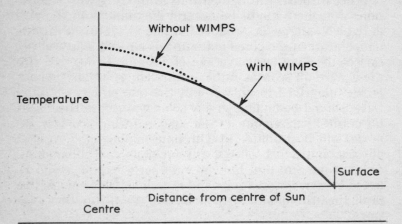

Figure 7.3/ The WIMPy Solution
If the Sun's core contains WIMPs, they will keep it just cool enough to explain why so few solar neutrinos are detected here on Earth.

a slightly lower temperature at the very centre, but keeping the outer parts of the core a little hotter than they "ought" to be.

How could the trick be worked? Faulkner and Gilliland proposed that there might be a kind of weakly interacting particle – a member of the WIMP family – which would be heavy enough to collect in the core of the Sun. There, a WIMP could collide with a hot proton, gaining energy itself and leaving the proton with less energy – that is, cooler. The WIMP particle, orbiting around the core, could then strike another, cooler proton, further out, and give up the energy it had borrowed. The process then repeats endlessly as the WIMP keeps moving around the solar core. The net effect is that the WIMPs take heat from the very centre of the Sun and distribute it more evenly through the core. WIMPs with the right mass and other properties (so that they interact just often enough with protons) can do this job so efficiently that relatively few of them, compared with the numbers of protons, are needed. But to do the trick you need just the right number of WIMPs (one for every 100 billion protons), with just the right mass, in just the right place. Faulkner and Gilliland found it easy to produce a model of the Sun with the same external appearance as the real Sun, and just the flux of neutrinos measured by Davis, with a core of constant temperature (an isothermal core) containing 6 per cent of the Sun's mass and kept isothermal by the presence of WIMPs. But they felt the whole thing was too contrived, and left their draft of the paper sitting on a shelf in Faulkner's office.

By 1985, however, WIMPs had become much more respectable, for reasons which had nothing to do with the solar neutrino problem. As neutrinos began to look un-suitable candidates for the dark matter in the Universe, and cold, dark matter models gained favour, cosmologists began to feel the need for WIMPs. At the same time, the particle theorists were developing their ideas of supersymmetry, which require the existence of WIMPs (see Appendix), and in addition other theories suggested that the Universe might contain a sea of stable axions. It was only a matter of time before someone thought of applying these now fashionable ideas to the solar neutrino problem.

When news came through to Santa Cruz on the astronomical grapevine that William Press and David Spergel, at the Harvard Smithsonian Center for Astrophysics, had come up with a way to keep the core of the Sun cool using CDM particles which they called "cosmions", metaphorical bells rang in Faulkner's mind. Gilliland, his PhD studies completed, had long since gone to work at the National Center for Atmospheric Research, in Boulder, Colorado; but Faulkner got the old draft paper down off his shelf and literally blew the dust from it. With a few very minor changes, to suggest how the Sun might gather up WIMPs as it ploughs round the Galaxy, the paper duly appeared, like the work by Press and Spergel, in the *Astrophysical Journal* in 1985.

The broad outlines of this new picture are that WIMPs (CDM) are at least part of the background of dark matter in the Universe, play a major part in the way galaxies form, and survive today as the material of the large halo of dark matter surrounding a galaxy like our own. The "right" cosmological distribution of WIMPs, and the appropriate properties from the particle physics, means that just one WIMP for every hundred billion protons is captured by the Sun, and because they are pulled in by gravity but otherwise only weakly interacting these particles orbit around in a cloud spread over about 10 per cent of the Sun's radius, interacting, on average, with one proton in each orbit. The WIMPs themselves would have masses of between 5 and 10 times the mass of a proton. Lawrence Hall, of Harvard University, produced a model in which the number of WIMPs in the Universe is exactly equal to the number of baryons; in that case, he calculates, a WIMP mass of 10 proton masses would make the Universe exactly closed, leave the proportion of baryonic mass at 10 per cent, fitting beautifully with estimates from the helium abundance, and solve the solar neutrino problem.

Hardly surprisingly, during 1985 and 1986, WIMP research began to become a thriving branch of astrophysics. On a visit to the Tata Institute in Bombay, Faulkner teamed up with Douglas Gough, a visitor from Cambridge, and M. N. Vahla, of the Institute. Gough had been involved for years in analysing small, regular changes in the surface of the Sun, oscillations as if it were breathing in and out. These oscillations are very hard to explain precisely in simple terms using the standard model of the Sun's interior, but Faulkner and Gough realised that the change in temperature gradient from the surface to the core required by the WIMP model alters the physical conditions in just the right way to make them match the simple physical theory. With Vahla's help, they calculated the details of the oscillations "by hand", as Faulkner proudly puts it, using old-fashioned paper and pencil.

Meanwhile, back in Colorado, Gilliland had latched on to the same idea. He had tackled the puzzle of solar oscillations using a large computer, and working in collaboration with Werner Däppen, also based in Boulder, and J. Christensen-Dalsgaard, from Aarhus University in Denmark. Once again, it turned out that the known solar oscillations could be best explained by a model in which the temperature at the heart of the Sun is set 10 per cent lower than in the standard models. Unknown to each other, working on separate continents, the two teams reached the same conclusion by independent routes, and their two papers were published alongside each other in Nature in May 1986. But even that was not the end of the story, for within a few weeks the WIMP model had passed the essential test of any good scientific theory – it had made a successful prediction.

The oscillations that were explained so neatly by the WIMP model are a kind of vibration called P-modes. These are pressure waves, which travel at the speed of sound through a fluid – rather like the waves that would be produced if you slapped the surface of your bath water with a flat hand. There is another mode of vibration, called G, or gravity, waves, which are equivalent to the waves that slosh up and down the bath when you sit in it. These move at slower speeds.

Faulkner and his colleagues predicted that there should be a G-mode oscillation of the Sun, if the WIMP model is correct, with a period of about 29 minutes. Standard theory predicts a fundamental oscillation with a period of about 36 minutes, but these waves are difficult to detect and had never been observed. Crucially, other tricks for reducing the temperature at the heart of the Sun, by tinkering with the astrophysics or the nuclear physics, force the period of these oscillations upwards, above 36 minutes. Within two weeks of the WIMP prediction being published in *Nature*, however, Faulkner heard that Claus Froehlich, of the World Radiation Centre in Switzerland, had found the predicted 29-minute period.

How could the Swiss have responded so soon? The evidence actually comes from old data, from a satellite known as the Solar Maximum Mission, dating back to 1980. Why had nobody spotted it before? Believing the prediction of the standard model, the observers had searched their data only for periods from a little below 36 minutes upwards. As Faulkner puts it, "The observers had too much faith in the theory." As soon as the Swiss team re-examined the old data, they found hints of the 29-minute period. These are, as yet, only hints, and a great deal more research needs to be done yet before the final triumph of the WIMP is confirmed. But, as things stand, particles five to ten times as massive as the proton, lurking at the heart of the Sun, can resolve the most embarrassing outstanding problem in astrophysics and account for 90 per cent or more of the mass of the Universe. These WIMPS have to be the "best buy" for the dark matter at the time of writing. The fact that we don't know which of the SUSY particles is involved is a strength of the model, not a weakness, since it can accommodate whichever flavour of WIMP the theorists prefer.

The initial success of the WIMP explanation of the solar neutrino puzzle has now encouraged experimental physicists to take the idea of WIMPs even more seriously than they might have on the strength of the theories alone, and, as we shall see, experiments are now being designed and constructed to trap the kind of particles that are now thought by some astronomers at least to permeate our Galaxy and to lurk at the heart of the Sun. Before we look at those practical possibilities for trapping WIMPs here on Earth, though, there are still some cosmological twists to the tale.

COSMOLOGICAL IMPLICATIONS

One of the reasons why some physicists prefer the name "cosmion" for the hypothetical particle needed to "explain" the solar neutrino problem is that this type of particle may not, in fact, be any of the ones actually proposed so far by the different particle theories. Perhaps the solar neutrino observations are actually pointing us towards something quite new. If so, it is remarkable how well the needs of the solar astrophysics and the need for dark matter on the scale of our Galaxy mesh together.

There is certainly no problem in explaining how the cosmions/WIMPs get captured by the Sun. If these particles exist in a stable halo around our Galaxy, then they must be moving around the galactic centre in their orbits, with velocities close to 300 km per second for those orbits to be stable. If such a particle is swept up by the Sun, whether or not it "sticks" inside the Sun depends on whether or not it has enough speed to escape the pull of the Sun's gravity. And the required escape velocity from the *centre* of the Sun is 3,000 km per second. Even at a distance out from the centre where half the Sun's mass lies "below" and half "above", the escape velocity is still 2,100 km per second. The implication is that a large number of the cosmions that pass through the surface of the Sun will indeed be trapped. But this very trapping causes some theoretical problems.

Cosmions in free space don't bump into one another very often, and it is quite possible that a sea of cosmions pervading the Universe could contain equal numbers of particles and antiparticles without all of them having been annihilated since the big bang. For a cloud of cosmions trapped inside the Sun though, interactions will be rather more common. They only interact weakly with *other* particles; when a particle WIMP meets an antiparticle WIMP, the result is still quick, and effective, annihilation. A balance will be struck between the rate at which new

cosmions (both particles and antiparticles) are gathered up
by the Sun, and the rate at which they are destroyed by
annihilations inside the Sun. The more cosmions and
anticosmions there are inside the Sun, the more likely they
are to meet one another and be eliminated. So the popu-
lation builds up only until the rate at which they are being
eliminated is the same as the rate at which they are being
swept up.

The limit on how many cosmions are gathered up is set
by the limits on the mass of dark matter in the halo determined from observations of the way galaxies rotate, and it
is very difficult to get the balance right (right, that is, for
the resolution of the solar neutrino problem) for any of the
familiar particles of supersymmetry and the other theories.
There are theories which allow the existence of particles
that do not annihilate so easily; and it is possible that
for some reason the Sun only sweeps up cosmions, not
anticosmions (or vice versa); but the most striking possibility is that the WIMPs inside the Sun – the cosmions –
were produced, like the baryons, with an excess of particles
over antiparticles in the big bang itself. This is the possibility implicit in the work by Lawrence Hall that I have
already mentioned; the implication is that if and when the
cosmion WIMP is identified, a study of its properties will
tell us about the conditions in the big bang itself, in much
the same way that the abundance of helium in the Universe today, and the presence of three flavours of neutrino,
is intimately connected with big bang physics. Cosmologists await the identification of the cosmion with
mixed feelings – they hope that it will fit the standard picture, but fear that its discovery may yet show up flaws in
the standard model of the big bang.

Hall himself calculates that if the dark matter survived
for the same reason that the baryons survived, because of
a kind of cosmic asymmetry, then the "natural" range for
cosmion WIMP masses is up to ten proton masses, which
is just right for the solar neutrino problem. This is certainly
enough to explain the dark matter needed for galactic
haloes, he says, but if the number of cosmions is equal to
the number of baryons, there are "not quite enough for
$\Omega = 1$". I don't think, however, that this need cause much

alarm. It might be tidy if we could explain all the missing mass in one package, but it seems far more likely that more than one of the possibilities raised by physicists in the past few years will turn out to have a bearing on the truth. Remember, for example, that the best theories now *require* neutrinos to have mass, but mass too small to close the Universe; a combination of neutrinos and WIMPs could be, like baby bear's porridge, "just right". *And although it might also be tidy to have the number of cosmions equal to the number of baryons, this itself is only a guess!

So researchers working at the interface between particle physics and cosmology have plenty of implications to ponder over if the WIMPy theory of the solar interior stands up to scrutiny. It might all seem more trouble than it is worth, if it were not for one last piece of information about the way the presence of WIMPs affects the nuclear fusion processes in the core of a star like the Sun. Faulkner himself had not yet worked through all the implications when I questioned him on the point late in 1986; he was just beginning to set up the computer calculations that would test the idea in detail. But the simple, preliminary calculations ("old-fashioned" pencil and paper, once again) tell us that if *all* stars contain a mixture of WIMPs in their hearts, and those WIMPs have the properties required to resolve the solar neutrino problem, then there will be a definite effect on the way the stars evolve as they age. Just as the outward appearance of our Sun is telling us, wrongly, that it "ought" to have a central temperature of 15 million degrees, so the outward appearances of stars have been used to calculate, perhaps wrongly, their *ages*. The size of the effect is what Faulkner is now working out; but the direction in which it operates is clear from the basic physics. On the WIMPy core picture, the oldest stars

---

* It should be no surprise, either, that all the different kinds of mass add up so closely to the critical density. If the inflationary scenarios that explain so well why the Universe is so smooth today are indeed correct, omega *must* be indistinguishably close to one. It is simply a question of how the available mass-energy got shared out in the big bang, and if a certain proportion went into WIMPs, say, then that could only leave so much available for baryons, or neutrinos, or whatever. Whichever way you slice the cake, the size of the whole cake stays the same.

in our Galaxy, blue stars in the globular clusters, must be significantly *younger* than they appear to be from the standard models of astrophysics.

Why should this be so exciting for cosmology? Because the accepted ages for these stars, 20 billion years or so, are a considerable embarrassment in the standard models of a closed Universe, with a value of about 50 for the Hubble parameter and an age of less than 15 billion years. Stars, obviously, must be younger than the Universe; WIMPs may explain how stars that look 20 billion years old can exist in a Universe that is only 15 billion years old – and may even give the leeway to push the Hubble parameter up a bit higher than 50, as many cosmologists would prefer. The solar neutrino problem, it seems, may be just about the *least* of the astronomical difficulties resolved by the WIMP theory – which makes it all the more important to capture one of these cosmions as soon as possible.

BRINGING IT ALL BACK HOME

By the cosmological standards of most of the subject matter of this book, the Sun lies in our own backyard, and bringing the search for the dark matter back into the Solar System is bringing it very close to home. But the latest generation of experiments planned by the physicists goes one better, bringing it all right back home, into the lab.

Davis's experiment is really the precursor for these new studies, the first of a family which are aimed at finding the traces of cosmions here on Earth. New generations of solar neutrino detectors are themselves in the advanced planning stage. If neutrinos do turn out to have mass, the experiments may turn out to be cosmion detectors, after all; whether or not neutrinos have enough mass to close the Universe, the new generation of solar neutrino detectors should be able to test the WIMP theory, by finding out whether or not oscillations are occurring and how many neutrinos the Sun is emitting at different energies. These experiments, however, are every bit as difficult and expensive as the example of Davis's detector would indicate. Davis himself, together with researchers from the Max Planck Institute, the Weizmann Institute, the Institute for Advanced Study in Princeton, *and* the University of Pennsylvania, has run a pilot study of a system using nearly one-and-a-half tonnes of gallium, in the form of a chloride, $GaCl_3$. A neutrino with an energy of only a quarter of a million eV can interact with gallium-71 to form germanium-71 and an electron; the idea is that the germanium is flushed out of the detector using a mixture of helium and hydrochloric acid, and the atoms counted. Tests suggest the scheme works – but a full-scale detector would use 50 tonnes of gallium, and there is no sign of funding for the experiment being available. The cost of the amount of gallium alone required for this experiment runs close up towards $10,000,000.

Other plans include: a study at the University of Oxford on using indium, cooled to below 3.4K so that it becomes superconducting, as a neutrino detector; a Canadian scheme involving a tank containing 1,000 tonnes of heavy water at the bottom of a mine in Sudbury, Ontario; and a detector using liquid argon tucked away in the Gran Sasso tunnel under the Alps. I mention them only to indicate that these are big, expensive and difficult projects. By comparison, as the experimenters have just realised, cosmions of the kind required to resolve the solar neutrino problem are *much* easier to detect. Their mass, the number of the particles in each cubic metre of space in our part of the Galaxy, their velocities and other properties should all be susceptible to test, without leaving the comfort of our laboratories even to go down a mine shaft or the Gran Sasso tunnel.

Once again, several experiments are planned, but they all make use of the simple fact that a cosmion WIMP is massive, more massive than a proton. If a particle like this hits an atomic nucleus, it stays hit. The "weakness" of the interaction that gives a WIMP its name means that they only interact occasionally with baryonic matter; if a WIMP does condescend to interact with an ordinary atomic nucleus then that nucleus certainly knows it has felt an impact, and recoils. One brief example of an archetypal cosmion detector – as yet still in the planning stage – should make it clear how physicists can turn this to advantage.

Researchers at the Rutherford Appleton Laboratory (RAL) in Oxfordshire are planning a detector designed to observe these galactic cosmions through their interactions with targets made of silicon and germanium compounds. The idea behind this is that the energy carried by a single one of these particles – simple kinetic energy, not mc^2 – is so great that when one strikes an atomic nucleus in a target of pure silicon one cubic millimetre in size it should raise the temperature of the target by five thousandths of a degree (5 mK). The target nucleus itself would recoil by 10 or 100 atomic diameters, shaking up its neighbours and sharing the energy of the cosmion among them as heat.

This temperature change is small, but it is big enough to be detected, in principle, using off-the-shelf low temperature technology. Indeed, single X-ray photons have already been detected in this way. Of course, there are problems to be overcome in putting the idea into practice. The detector has to be shielded from the "noise" of stray X-rays, ordinary cosmic ray particles, and the products of any radioactive decays occurring in impurities in the detector itself. But this is all in the day's work for the experimental physicist, and the technique has the great advantage that a detector using only one kilogram of target material* (instead of the many tonnes for neutrino detectors) should find one cosmion per day if the WIMP theory stands up. Such a relatively cheap, straightforward laboratory experiment could locate the missing mass *and* check out the theories of supersymmetry, while the neutrino experimenters and the physicists who specialise in colliding high-energy beams of particles together are still struggling to find enough money to make the

---

* Not all in one lump, of course, but as an array of tiny cubes each one wired, not for sound, but for temperature.

next step forward with their own lines of attack on the problem.

And that is as far as the search for the missing mass can take us as yet. We have travelled to the farthest reaches of the Universe and the beginning of time itself, only to be led back to studies being done here and now, studies that may, within a few years, produce traces of dark matter before our very eyes, if not quite in the palms of our hands. There is no room to doubt that there is *something* out there, although we do not yet know just what it is, or how much there is. What we see is no more than 10 per cent of the Universe. But is the Universe open or closed? What is its ultimate fate to be? There is still, just, room for theorists to speculate, offering a choice of future possibilities for us to wonder at.

CHAPTER EIGHT

A CHOICE OF FUTURES

One sure thing about the ultimate fate of the Universe is that it lies a long way ahead, by any human time-scale. It is not something we have to worry about in any practical sense, in the way that we should (if we were sensible) worry at least a little bit about when the next Ice Age is coming here on Earth, or when the oil supplies on which our global society is so dependent will begin to run out. For a human being, the farthest we can peer into the future to discover an ending which might bring a slight tingle up the spine is about five billion years, to the time when our Sun will swell up to become a red giant star, and will engulf the Earth in the process. The death of our own home planet seems like a pretty final full stop to write, to us. But this represents scarcely a hiccup on the cosmic scale of things, whatever the ultimate fate of the Universe itself. In both open and closed scenarios, matter will be a long time a-dying.

THE FATES OF STARS

For the moment, let's set aside the new discoveries that tell us that 90 per cent of the Universe is not composed of familiar baryons in the form of stars and planets. What will happen to the kind of matter we do know and love as the Universe ages? Stars themselves, though they may burn nuclear fuel to hold themselves up against the pull of gravity for billions of years, cannot last forever. When their fuel is exhausted, gravity must win its long, drawn-out battle, and cause the stellar remnants to collapse. And, it turns out, there are only three things that a dead star can collapse into.

Stars form from clouds of dust and gas in space, by processes which are still not very well understood by physicists. But it is very clear how they get hot, and stay hot, once they have formed. Gravitational collapse releases heat which makes the young star glow brightly, and as the star collapses and becomes more compact and dense that heat is enough to initiate fusion reactions in its heart, like the ones which keep the Sun hot today, and which ought to be producing copious floods of neutrinos for Ray Davis to detect. The more massive a star is, the more fuel it must burn each second to hold itself up against the pull of gravity. Some stars exist in this stable main sequence state for only a few million years; the Sun has already done so for 4.5 billion years, and has about as long still to go before its hydrogen fuel is exhausted; other stars may live for even longer. Things must change, though, when all the hydrogen in the core of a star has been converted into helium.

With no more hydrogen to burn, the star is no longer able to resist the pull of gravity, and the core begins to collapse once more. But this releases more gravitational energy as heat and new nuclear reactions begin, as a result, converting helium into carbon. The energy released in the process makes the outer layers of the star expand, and this is why the Earth will be engulfed by the Sun in about five billion years from now. Eventually, the helium will all be converted into carbon, and the process will repeat as before, with the core getting more compact and hotter still while carbon is converted into oxygen by fusion. The process can repeat several times, until most of the core material has been converted into iron-56. But there it must end, because the addition of more protons and neutrons to the nucleus of iron-56 does not release energy. Instead, in order to build up heavier elements energy must be put *in* from somewhere.

In some large stars, that is exactly what happens next. With its source of heat at last removed, the star collapses, with the mass of the outer layers – its distended atmosphere – falling inwards under the pull of gravity, and themselves getting very hot as gravitational energy is released. In these conditions, given enough mass, there can be a sudden explosion of fusion activity as hydrogen and helium from the atmosphere of the star squeeze down on to the core. The explosion takes place in a shell around the core, like the peel surrounding an orange, and the resulting blast travels in two directions – outwards, ejecting the rest of the atmosphere of the star to form a glowing, expanding nebula, and inwards, squeezing the core tight and also producing a smattering of elements heavier than iron, some of which may get thrown out into the new nebula. The star has become a nova, or a supernova. In its death throes, it has helped to seed the interstellar medium with heavy elements that will go into the next generation of stars and planets. It is only because previous generations of stars have gone through this cycle that our Sun has a family of planets at all, and we are here, with our bodies based on carbon chemistry and breathing oxygen out of the air, to puzzle over the origins of clouds in space like the famous Crab Nebula.

Not all stars, however, end their active lives in such spectacular fashion. Less massive stars, like our Sun, may fade away more quietly, puffing out a little gas into space but then settling down gently as glowing cinders once the processes of nuclear fusion are exhausted. Just what they settle into depends on their mass.

A dying star, on this picture, consists of a dense ball of matter, chiefly in the form of nuclei of iron swimming in a sea of free electrons. Of course, there will still be a thin atmosphere of hydrogen and helium, and probably a crust rich in nuclei such as those of carbon, but we can ignore those. The most important particles, as far as the first stage of star-death are concerned, turn out to be the electrons.

Electrons have a very important property, which they share with other entities that we are used to thinking of as particles, such as protons and neutrons. This family of particles is known as fermions, after the Italian physicist Enrico Fermi, and no two fermions can ever share exactly the same quantum "state". This is the reason why the electrons in an atom, for example, are spaced out around the nucleus. When physicists first discovered that electrons in atoms formed a cloud around the nucleus, they were greatly puzzled, because the electrons have negative charge and the nucleus positive charge, and opposite charges, of course, attract. Why didn't the electrons all fall in to the nucleus? In essence, the answer is that if they did so they would all be in the same state, in the same energy level. In order to keep its own uniqueness, each electron associated with an atomic nucleus has to find its own place, somewhere near the nucleus, jostling with the other electrons that belong to that atom in a cloud surrounding the nucleus.

There are "particles" which are not fermions, and which are called bosons, after another pioneering physicist, the Indian Satyendra Bose. These are the particles that we often think of as waves, or radiation. Photons, for example, are bosons, and they are quite happy to crowd into the same state as one another. Indeed, this is the basic principle of the laser – the powerful beam of light which a laser emits is made up of countless numbers of photons, all in exactly the same state, all marching in step with one another. But that has nothing to do with what holds a dead star up against the pull of gravity.

As a dead star cools and shrinks, there comes a time when all of the electrons are packed so closely together that they occupy all of the possible states that exist for them inside the star. They can be squeezed together no more, since that would involve forcing several electrons

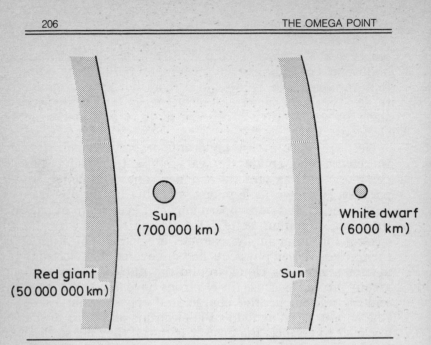

Figure 8.1 / Sizes of Stars
The range of possible stellar sizes, shown to scale.
Planet Earth is roughly the same size as a white dwarf;
a neutron star is the size of a mountain.

into the same state, and they resist the inward tug of gravity with an outward pressure, called electron degeneracy pressure. Provided the mass of the stellar remnant is less than about 1.4 times the mass of our Sun, this is the end of the story. Gravitational force balances the electron degeneracy pressure when a star like the Sun has shrunk to about the size of the Earth, and it is called a white dwarf. Astronomers know many white dwarfs, and, indeed, such stars were discovered before the theory I have outlined here was put together to explain their origins. As such a star cools, it stays much the same size, but gets dimmer and dimmer, fading through the brown dwarf stage to become, ultimately, a black dwarf – a ball of iron, surrounded by a shell of carbon, and perhaps even a trace of ice from the remnants of oxygen and hydrogen in its old atmosphere. The fate of our Sun is to become a sooty ball of ice-streaked rust.

Neutron star
(10 km)

Neutron star

Black hole
(3 km)

White
dwarf

But what if the mass of the star is more than 1.4 times the mass of our Sun, even after it has blown away its outer layers in a last blaze of glory? Although the electrons still cannot occupy the same state as each other, under such intense gravitational pressure they find another escape route, as, in effect, electrons are squeezed into protons to form neutrons, releasing neutrinos in the process. It might seem that they would be happy to do so, since positive and negative charges are attracted to one another, but in fact there are other forces at work in atomic nuclei (and under comparable conditions), which keep protons and electrons apart unless the squeeze is overwhelming. I won't go into the details here; what matters is that if a dead star has got more than 1.4 times the solar mass of material doing the gravitational squeezing, all the electrons and protons it contains are converted into neutrons, and it collapses a stage further, to become a ball of neutrons – in effect, a single "atomic" nucleus – about 10 kilometres in radius. It has become a neutron star. At this point, it is held up by the degeneracy pressure of the neutrons, exactly similar to the electron degeneracy pressure which holds a white dwarf up against the pull of gravity. But this pressure can only resist gravity if the mass of the stellar remnant is less than about three times the mass of our Sun.

Some physicists have suggested that there may be a stage beyond the neutron star stage, in which the neutrons are reduced to their ultimate components, quarks, and the star is made of quark "soup". But this has very little effect on the calculations, since quark soup is just about the same density as neutron matter, anyway. With no nuclear fuel to burn, any dead star with a mass of more than three times the mass of our Sun cannot hold out against gravity at all, but must collapse into the ultimate sink, a black hole, while the matter it once contained is literally crushed out of existence. Neutron stars and black holes may very well be made in the explosions that produce supernovae; this would also be a way to make neutron stars with less than 1.4 solar masses, by squeezing their material together at the hearts of exploding stars. Black holes will also be formed if another compact object – a neutron star, perhaps – sweeps up enough matter from its surroundings to exceed the three solar mass limit. Whatever their origin, there are plenty of these compact objects about. Hundreds of neutron stars have been detected, in the form of pulsars, and several likely black holes are known. Every star *must* end up as a white dwarf, a neutron star, or a black hole. But what happens then?

THE FATE OF MATTER

In round terms, the age of the Universe today (the time that has elapsed since the big bang) is about 10 billion years, or 10^{10} years. Cosmologists can calculate how matter will evolve as time passes in model universes with different sizes. The best bet, as we have seen, is that our Universe sits very close to the dividing line between being open and closed, with omega very close to one. In that case, it is likely to continue to expand for a very long time indeed, before it turns around, ever so gradually, and begins to shrink back towards the omega point. In constructing such cosmological models you can make the cycle time of the Universe as long as you like, by making omega bigger than one but setting it as close to one as you like, without ever allowing it to be exactly one – or, of course, you can make the lifetime of the model universe *infinitely* long, by setting omega just below (or a lot below!) one. The fate of matter depends on how long the Universe is around for different processes to work themselves out before the big crunch.

Just about the smallest time-scale worth considering is for a closed universe which begins to contract about 10^{11} years after the big bang. That is, in a time ten times longer than the present age of our own Universe. Nobody seriously suggests that we live in a universe that small, but it is just about possible to reconcile observations of the real Universe with the requirements imposed by such a model. In a universe this small, the same sort of processes of star formation, planet formation and – presumably – the emergence of life will still be going on in galaxies like our own even after the time the universe turns around and begins to collapse. Leaving aside the philosophical questions about the meaning of time in the collapsing phase of the universe, the first landmark event after the turn around will be when the universe has shrunk to one-hundredth of the size of our Universe today, when galaxies begin to merge into one another. Life as we know it might be around even under those conditions, and would measure a background radiation of only about 100 K. At a size one-thousandth of that of the Universe today, however, the sky has become as bright as the surface of our Sun because of the blueshift effect, piling up radiation from earlier epochs; the "background" radiation has a temperature of 1,000 degrees, and life as we know it is impossible.

At one-millionth of the present size of the Universe, stars explode as the background temperature reaches several million degrees, comparable to the temperature inside the Sun today; at a billionth the present size, nuclei are broken up into protons and neutrons at a temperature of a billion degrees; and at one-trillionth of the present size protons and neutrons are smashed apart into a quark soup, with a temperature of about a trillion (10^{12}) degrees.

It is certainly a dramatic image of the big crunch,* but one that should be treated with caution. First, our Universe contains ten (or more) times more dark matter than it does baryons, and we cannot be sure how these particles will affect the baryons during the collapse. Secondly, there is no evidence that our Universe, even if closed, is small enough for this scenario to play itself out. Indeed, all the evidence is that our Universe is either just open or (the interpretation I find most persuasive) only just closed. In either case, there will be ample time available for very long-term, quantum effects to come into play. And that gives matter as we know it a very different ending.

If the expansion of the Universe continues for long enough, star formation will cease as all but a trace of the available hydrogen and helium is used up. From studies of the distribution of old and young stars in our Galaxy, and calculations of the rate at which star-making material is being used up, astrophysicists estimate that this will happen in about a trillion (10^{12}) years from now. Galaxies will become redder, as their stars age and cool, and then, eventually, fade away as all the stars they contain become white (ultimately black) dwarfs, neutron stars or black holes. Over very long time-scales, galaxies will shrink. This is partly because they lose energy through gravitational radiation, and partly a result of inevitable encounters between stars in which one star gains energy and is ejected from the galaxy while the other loses energy and falls towards the centre of the galaxy. In a similar fashion, clusters of galaxies will shrink in upon themselves, and eventually both individual galaxies and clusters will fall in to huge black holes of their own creation.

It is difficult to come to grips with what "eventually" means. The numbers cosmologists play with look simple enough – 10^{15} years, 10^{20} years, and so on. But remember that each additional unit on the "power of ten" means *multiplying* tenfold. The age of the Universe is about 10^{10}

---

* Borrowed, in part, from the excellent (but weighty) book *The Anthropic Cosmological Principle*, by John Barrow and Frank Tipler.

years; 10^{11} years is *ten times longer* than all the time that has elapsed since the big bang. Similarly, a trillion years (10^{12}) seems mind-bogglingly long to us, but 10^{15} years is a *thousand* times longer still; and 10^{20} years isn't twice as long as the age of the Universe, but ten billion times as long! Even this, however, is a mere eyeblink compared with the time-scales required for the ultimate decay of matter.

According to some of the most favoured present-day theories of particle physics (which are favoured because they have been proved right so many times already), protons themselves should be unstable, and must decay, each one turning into a positron, a shower of neutrinos and gamma rays (neutrons inside a white dwarf or neutron star do much the same, but produce an electron as well as a positron, keeping the balance of electric charge). The same rules which allow baryons to be produced in the big bang suggest that they must ultimately leave the cosmic stage. But the time-scale required for this process is very long indeed. In a lump of matter (any lump of baryons), half of the protons will decay in rather more than 10^{31} years. For everyday purposes, the proton is very stable – which is just as well, or we wouldn't be here. But for a slowly cooling white dwarf, proton decay becomes important. Without proton decay, a white dwarf will radiate all its heat away and become a black dwarf at the same temperature as the background radiation in about 10^{20} years. Protons decaying inside it, however, can provide enough energy to keep it as warm as $5\,K$ (a mere -268 C) until 10^{31} years have passed. That doesn't sound too hot, but at that time the temperature of the cosmic background radiation will be only $10^{-13}\,K$, which puts it in a more impressive perspective. Neutron stars, being more compact, are kept hotter, perhaps as warm as $100\,K$, for the same time. By then, half the baryons have been used up.* But by 10^{32} years, virtually *all* the baryons have gone, because, of course, 10^{32} is *ten times* more than 10^{31}.

By this time, all objects made of baryons have lost almost

---

* Assuming, for the sake of argument, a proton "half-life" of *exactly* 10^{31} years. This is less than the limits set by the latest experiments, but the argument is the same whatever the exact numbers you put in.

all their mass. Black dwarf stars have shrunk to the mass of the Earth, and a planet like Earth will have shrunk to the size of an asteroid. By the time 10^{33} years have passed, all the baryons in the universe will have gone, converted into energy, neutrinos, electrons and positrons. Today, every proton in the Universe is balanced by an electron, so that there is no overall electric charge. After this enormous time has passed, and baryons have decayed, the remaining form of matter will be equal numbers of electrons and positrons, scattered across the universe. When lumps of matter, like stars, decay in this way, the positrons and electrons will quickly meet one another and be annihilated, releasing more energy in the form of gamma rays. But perhaps as much as 1 per cent of all the original baryons in the universe will still be in the form of hydrogen gas after star formation has stopped, and when the nuclei of these isolated atoms of hydrogen decay the resulting positrons can pair up with the electrons of the original atoms, orbiting one another at a safe distance in a kind of pseudo-atom called positronium.

After 10^{34} years, the universe will contain nothing but radiation, black holes and positronium. Even a black hole, however, is not forever. The Hawking effect will slowly convert black holes themselves into particles and radiation as they "evaporate". A black hole with the mass of a galaxy will evaporate in 10^{99} years, and even a hole containing the mass of a supercluster of galaxies – the biggest likely to form – will be gone in 10^{117} years. The ultimate products of this evaporation will be more electrons and positrons, more neutrinos, and more gamma ray photons. So, after 10^{118} years, if the universe lasts that long, we come to the ultimate fate of matter – to be converted into positronium, neutrinos and photons. And if the GUTs are wrong and protons do not decay as expected, this only shifts the time-scale a little matter of four powers of ten, since even a proton will evaporate through the Hawking process after 10^{122} years!

In a closed, but long-lived, universe, recollapse eventually occurs and the plummet towards the omega point still happens, but there are no stars and galaxies to be disrupted in the process. Just a broth of electrons, positrons, neutrinos and photons being squeezed into the ultimate singularity. Somewhere in between the two extremes, medium-sized closed universes will recollapse while they still contain a mixture of dead stars in one of their three guises and some gas and dust that has not been processed into stars.

It is pure guesswork how the dark matter will affect these calculations, since we do not know what the dark matter is. Even so, it seems that those who like to speculate about these things have a choice of futures to "believe in", and can pick the one that they find most comforting. But this is still not quite the end of the story, since there is another sense in which we seem to have a choice of universes, an interpretation of cosmology which brings us, like a closed universe, round full circle and back to the more philosophical considerations with which this book began.

A CHOICE OF UNIVERSES

The most successful description of the physical world that we have is the quantum theory, which tells us how particles and forces behave on the scale of atoms and smaller. Even the general theory of relativity (though it *has* passed every test it has been set) has not been tested as fully as the quantum theory, which is the physicists' *pièce de résistance*. But this theory tells us some very strange things about the nature of reality, strange things which *have* to be accepted as the literal truth if the successes of the theory – which range from lasers to molecular biology to high-speed electronic computers – are to be explained at all. Perhaps the oddest feature of the theory is the interpretation of what it actually means to say that a particle such as an electron is in a particular quantum "state".

I won't go into all the gory details here – I did that in my book *In Search of Schrödinger's Cat*. But one simple example will help to highlight this quantum oddity. Every electron carries a property which physicists choose to call

spin. You can think of it, if you like, as an arrow, or a
spear, which is carried by an electron and can only point
in one of two directions, "up" or "down". Really, we
shouldn't try to think of these quantum properties in
everyday terms at all, but it is the only way to get any
kind of a picture of what is going on. Whatever the image
you carry in your head, though, what matters is that an
electron with spin up is in a different state from an electron
with spin down. This is why two electrons can both crowd
into the lowest energy level available in the helium atom,
for example, without being in the same state. Because one
has spin up and one spin down they can share the energy
level, in a sense both at the same distance from the atom's
nucleus. If they both had the same spin, this would be
excluded, because they are fermions. In a more complex
atom, the next electron has to go into a higher energy
level, further out from the nucleus, because whichever
possible spin it has it is excluded from the lowest level by
the presence of another electron with the same spin – but
that requirement, with all that it implies for chemistry, is
not the story I want to go into here. Instead, imagine a
single electron, sitting on its own, or travelling through
space. What spin does it have?

Now, we can do experiments to measure the spin of the
electron, and when we do we will always find that it either
has spin up or it has spin down. But the quantum theory
tells us that when the electron is left to its own devices, it
neither has spin up *nor* spin down, but exists instead in
some mixture of both possibilities, called a superposition of
states. The "reality" of an electron which has "collapsed"
into a single, definite spin state exists only when it is being
measured, or when the electron is interacting with another
particle. Once the measurement (or interaction) has
ceased, the unique spin state dissolves away once more
into a superposition of states. At this level, things only have
a unique, "real" existence when they are being looked at or
prodded in some way. And this bizarre behaviour holds
true for *all* quantum properties, not just spin; it is an *essential* feature of quantum physics. Without it, we would not
be able to explain how lasers work, or why DNA forms a
stable double helix, or how semiconductor chips do their

tricks inside computers, to name but a few. As for what this strange behaviour implies about the nature of reality, physicists and philosophers have been debating the issue, on and off, for half a century. Now, the cosmologists have got in on the act.

There are two principal ideas about what the collapse of an electron (or anything else) from a superposition of states really means. The conventional wisdom is called the "Copenhagen Interpretation", because a lot of pioneering work on quantum theory was done at the Institute founded by Niels Bohr in Copenhagen. In fact, a key ingredient of this interpretation actually came from Germany, from the work of Max Born. This is the idea that the behaviour of things at the quantum level is ruled by chance – not in the whimsical, fluky sense that we sometimes imagine our lives to be ruled by chance, but in the sense that the behaviour of electrons and the like follows the strict statistical rules of probability – the rules the Casino takes good care to apply to make sure it keeps an edge. In terms of our isolated electron, this means that when its spin state is measured there will be a precise 50:50 chance of finding it with spin up, and the same chance of finding it with spin down. Measure a million electrons, or the same electron a million times, and half a million times you will get the "answer" spin up, half a million times you will be told spin down. But you can never predict in advance what the outcome of any one of the individual measurements will be, only the relative probabilities of all the possible different outcomes. The same thing happens when you toss a coin. It has a 50:50 chance of coming up heads (assuming the coin is properly balanced), and each time you toss the chance is the same, even if you have just had a string of heads or tails. For an individual electron, even if you measure its spin and get the answer "up", there is only a 50:50 chance that the next measurement will give the same answer. It is only because real-life experiments involve huge numbers of quantum objects, all following the rules of probability, that their *overall* behaviour can be predicted, for practical purposes, using statistics. Half the electrons have spin up, half spin down, and a TV tube, for example, doesn't care which half are oriented which way. The quantum rules of probability work rather like the way in which life insurance companies make their money – they cannot tell in advance *which* of their individual clients will die in any particular year, but they do know from their actuarial (statistical) tables, *how many* will die, and budget accordingly.

The choice of spin states for a single electron is a very simple example, and real quantum systems are much more complicated superpositions of states, governed by suitably more tricky rules of probability. The probabilities themselves may be changed by the measurement, or interaction. For example, one of the other strange features of quantum physics is that an object like an electron is not confined to a definite location. It has a certain probability, which can be calculated, of turning up anywhere at all (this is related to the wave-like aspect of the electron's dual nature). The probability is very large that the electron will be found somewhere near where you last saw it, but there is a real, if tiny, probability that it will turn up somewhere else entirely. When you actually do measure the position of an electron, these probabilities all collapse into the certainty that it is where you saw it. Once you stop looking, however, it once again has an opportunity to be somewhere else. Looking at the electron changes its probabilities, and alters the superposition of states the electron entity is in.

It sounds crazy. But the strangest thing about the Copenhagen Interpretation is that it works perfectly as a tool to describe what will happen as a result of our interference with the quantum world. It is the tool used by people working at the practical level, and it has proved its practical value repeatedly. The equations work. But nobody has the faintest idea what the interpretation really means, in terms of what electrons and things "do" when nobody is looking at them. This has provided scope for an alternative interpretation to be suggested, one which gives exactly the same "answers" as the Copenhagen Interpretation in all practical applications, but which has a different philosophical basis. It stems from the work of the American Hugh Everett in the 1950s; for reasons which will soon be apparent, this is called the "Many Worlds Interpretation".

QUANTUM COSMOLOGY

The Many Worlds version of quantum theory says that when you measure the spin of an electron it does *not* collapse into one spin state (chosen by the laws of probability) while you are looking at it, and then revert to a superposition of states. Instead, according to this interpretation, the world splits into two separate realities, in one of which the electron has spin up, and in the other of which the electron has spin down. The two worlds then go their separate ways, nevermore to interact with one another. Physicists and mathematicians are still arguing about exactly what this means — especially when scaled up to cover systems with more complex superpositions of states than an electron with only a choice of two spins. The fundamental thing to grasp, though, is that calculations carried through on the basis of the Many Worlds Interpretation give exactly the same answers as the Copenhagen Interpretation, always, when applied to practical problems. So it is just as good a tool (no better, no worse) for designing computers, or lasers, or in calculating the chemistry of complex molecules. The implications, though, are more than interesting.

One idea is that the whole universe splits into two or more replicas of itself every time any quantum system is forced to choose between possible states. This has led philosophers to fling up their hands in horror, unwilling to accept the implication that my measurement of the spin of an electron, here in a lab on Earth, can affect galaxies and quasars millions of light years away, instantaneously, as the entire universe splits in two. Or, indeed, that every time a quantum "choice" is made in some distant quasar, you and I, and everyone here on Earth, is duplicated into a myriad copies. (Science fiction writers, of course, love the idea!) But there is a way to appease the philosophers. Researchers such as Frank Tipler, of Tulane University, prefer to associate the "splitting" only with the quantum system involved in the interaction, or the experimental apparatus involved in making the quantum measurement. "There is only one Universe," says Tipler, "but small parts of it – measuring apparata – split into several pieces."*

Now, all this involves some fairly hairy mathematics, routine for the experts but a little over my head, and some unfamiliar philosophical ideas. This is current research, which is not yet cut and dried. But the implications of applying Many Worlds theory to cosmology are so startling that it may yet supplant the Copenhagen Interpretation, just as, in the seventeenth century, the Copernican interpretation of planetary motions in terms of the Earth going round the Sun supplanted the Earth-centred ideas of Ptolemy. It also rounds out the theme of this book beautifully. The idea is to take the equations describing the growth of the Universe from the big bang and to treat them in quantum terms using the Many Worlds Interpretation. On this picture, the Universe is split into many different branches by quantum processes in the beginning, at the moment of creation when the "size" of the Universe, as far as that has any meaning, is the size of a quantum fluctuation. These branches will have different properties from one another, but will be part of a single family, governed by a single set of rules. From the point of view of the present book, the most important feature is that there will be branches which have all permissible values of omega. That might seem like a disadvantage of the interpretation, since it seems to beg the question of why omega is so close to one in our Universe. But it turns out that this special value of omega is a *requirement* of Many Worlds cosmology.

This approach to cosmology was pioneered by Jayant Narlikar and his colleagues at the Tata Institute in Bombay, in the early 1980s. The work by Stephen Hawking and Jim Hartle, that I mentioned in Chapter Three, is part of the same approach to cosmology. Hawking's description of the Universe in terms of a closed spacetime, with no edges and no beginning or end, derives from the Many Worlds Interpretation. And Tipler, using a

---

* The quote comes from a paper by Tipler in *Physics Reports*; he covers the same ground in part of Chapter 7 of his book with John Barrow.

slightly different mathematical approach, comes up with the same key conclusion: quantum gravity combined with Many Worlds cosmology leads inevitably to the prediction that $\Omega = 1$. Many different universes, with many different sizes, can exist in principle, but in practice it is overwhelmingly likely that uniform, isotropic universes with exactly the closure density will be produced out of the big bang. Enthusiasts for this idea, such as Tipler, stress that this is independent of the idea of inflation, and that Many Worlds cosmology requires that omega be *indistinguishably* close to 1, while inflation "only" requires that omega is between 0.999999 and 1.000001 – not that anyone is ever going to measure it to this accuracy!

This is just one example of current thinking on the nature of the Universe, its beginning and its ultimate fate. Cosmology today is one of the liveliest branches of science, with ideas in ferment, and scenes shifting from month to month and year to year. Some ideas will fall by the wayside; others will stand the test of time. Nobody can yet be sure just how the detailed picture will look, especially regarding the fate of the Universe, in five, or ten, years' time. But I choose to end with this particular scenario for one very good reason. If that *is* the way the Universe is, then we are left with the best (or worst) of both alternatives. Ultimate collapse back into the omega point is assured, but the turn around will happen so slowly that there will be ample time, after all, for matter to decay. That is the neatest conclusion I can offer to this particular book, since the scientists themselves are still debating the issues, and in particular they are still puzzling over exactly what form the dark matter might take. The best evidence, certainly, is that the Universe is closed, and its ultimate fate is assured – but the search for the missing mass goes on. I'll let you know how it turns out.

APPENDIX

SUSY, GUTS AND STRING

The best way to gain some insight into the way physicists believe the world is constructed is to start out from the ancient history of particle physics – which means just over twenty years ago, in the middle of the 1960s. At that time, the material world was seen as made up of two kinds of particle. First, there were the leptons, particles like the electron and its neutrino, which do not feel the "strong" force of nature, and which have zero size, as far as any experiment has been able to test. We now know of six leptons (the electron, muon and tau particles, and their associated neutrinos), and they are still regarded as truly "fundamental". All the other known particles, including protons, neutrons, and hundreds of others, were called hadrons. They do feel the strong force of nature, and they are complex objects with a definite size.

Interactions between these particles were described in terms of four forces of nature. These are gravity, the electromagnetic force, and the strong and weak nuclear forces. Over the past two decades, particle physics has progressed towards an understanding of what hadrons are made of, and towards a description of the four forces of nature in one set of mathematical equations. A key step along this path is the development of theories that attempt to describe all the forces except gravity in one package; these are known as Grand Unified Theories, or GUTs.

The modern version of particle physics has produced a "standard model" which is used as a working basis for calculations and for describing what is going on in the particle world. Hadrons are now thought to be composed of quarks, which, like leptons, are truly fundamental particles with zero size. There are six types of quark, just as there are six types of lepton, and this balance between the two families is regarded as a significant confirmation of the validity of the standard model. The electromagnetic force acts between charged particles, and is very well described, for practical purposes, by a set of equations which incorporate the rules of quantum physics, and are known as quantum electrodynamics, or QED. The strong force acts between quarks, and is described by a set of equations, developed partly by analogy with QED, which are called quantum chromodynamics, or QCD. The "chromo" part of the name comes from the arbitrary decision of physicists to label the equivalent of "charge" for the strong force with the names of colours; but this does not mean that quarks are "coloured" in the everyday sense of the word.

Mathematical physicists have now found a way to combine QED and the description of the weak force in one set of equations, as the "electroweak" force. This development is based on the idea that at high enough energies – high enough temperatures – the two forces are indistinguishable, and that the symmetry between them is only broken at lower temperatures. Extending this idea, it is thought that in the first instant of the birth of the Universe as we know it, *all* the four forces were equivalent to one another, and describable by a super-GUT. As the Universe expanded and cooled, each of the forces split off in its own right at the appropriate temperature.

This idea can be understood in terms of the modern description of forces as being carried by an exchange of particles. The often used analogy is with a pair of skaters on a broad, frozen lake, who throw a heavy ball back and forth. The recoil as a skater throws the ball, or the impact as the partner catches it, forces them apart. The analogy does not explain how forces of attraction work, but perhaps it helps a little.

In the particle world, the exchange particles of the force need not have a "real" existence. They can be conjured up out of the vacuum, provided they only live for a short time, determined by Heisenberg's uncertainty relation, a fundamental feature of quantum physics. The range of such a "virtual" particle depends on its mass, because the more massive it is the less time it is allowed to exist. It has to be created, exchanged, and disappear all within the time allowed by quantum uncertainty. The weak force has a very short range, because it is carried by very heavy particles, called W and Z. Electromagnetism, on the other hand, is carried by photons, which have zero mass. A zero mass "virtual" particle can live forever, so the range of the electromagnetic force is, in principle, infinite.

What does this tell us about the breaking of unification? In the very early Universe, when there was plenty of energy about, there would also have been many W and Z particles, for example, existing in their own right. If the energy density of the Universe were higher than the energy needed to maintain a sea of W and Z particles, then they would not be subjected to the constraints imposed today by the uncertainty limit. Like photons, they would have infinite range, and there would be no way to distinguish between the electromagnetic and weak forces.

In the same way, physicists believe that at high enough energies all four forces were the same. Indeed, the only mathematically satisfactory descriptions of the electromagnetic and weak forces in one "electroweak" package are as part of much bigger schemes, GUTs, which claim to include the strong force unification as well. Only the electroweak part of these theories can yet be tested, however – it predicts the masses of the W and Z particles (about 100 times the mass of the proton), and these have been measured in the European particle physics laboratory, CERN, in Geneva. It takes enormous energies to manufacture such particles, and no accelerator on Earth can produce enough energy to measure the further predictions of the GUTs, appropriate for further stages of the unification. This is why particle physicists are now so interested in cosmological theories of the early Universe, and in the search for dark matter remnants of those early, high energy conditions.

But there are problems with all these ideas. One problem is that there are several different versions of Grand Unified Theory, which clearly says that the mathematicians have not yet found a single underlying truth. Another, crucially important, is that although theories like QED and QCD work at one level, they have only been obtained at all by some dubious mathematical trickery. In simple terms, the equations are plagued by infinities, which arise naturally and which can only be got rid of by dividing one infinity by another – a practice which we were all taught in school, and rightly, is forbidden. And it has proved very difficult to include gravity in the same scheme as the strong, weak, and electromagnetic forces. What physicists need – their Holy Grail – is a theory that includes gravity with all the other forces, is unique, and which is not plagued by infinities. In the middle of the 1980s, the world of particle physicists is in ferment because of hints that they may be very close to finding such a theory.

The first step along this road was taken with the development, in the mid-1970s, of an idea called supersymmetry, usually abbreviated as SUSY. The idea is as grand as its name suggests – it aims to unify not just the four forces but the material world as well in one mathematical description. This seems absurd – forces are forces, and particles are particles. How can a proton be described in the same way as a photon? The problem seems even worse when cast in terms of quantum physics, because the particles of the material world and the carriers of force form two very distinct families. Leptons, quarks and the particles made up out of combinations of these fundamental particles are all members of a family called fermions. Their distinguishing feature is that they all have a property called spin, which comes in half-integral units. That is, a fermion may have spin $1/2$, $3/2$, $5/2$ and so on, but never spin 1, 2, 3 or any other integer. Force carriers, on the other hand, are all bosons, and either have zero spin (like the photon) or spin 1, 2, 3 or some other integer value.

This is just a kind of quantum labelling of the particles, but it relates to a profound difference in properties. Fermions, such as the electron, keep themselves to themselves, with no two identical particles sharing the same quantum state. Bosons, on the other hand, swarm together. The two families are far more different than the proverbial chalk and cheese. But according to supersymmetry theory, for every type of boson there is an equivalent fermion, and for every type of fermion there is an equivalent boson. Since physicists already knew, by the time this idea was aired, that no *known* fermion could be the partner of any *known* boson, and vice versa, this meant that the theory required, at a stroke, doubling the number of possible types of particle in the Universe. If electrons exist, then so must selectrons; if quarks exist, so must squarks; photons are mirrored in the fermion world by photinos; W particles by Winos; and so on. Where, if supersymmetry was such a good theory, were these myriads of "new" particles to be found?

A few of these supposed superparticles, including the photino, might have quite low masses and might not have yet been detected because they interact only weakly with everyday matter. Sounds familiar? As I said, today particle physicists are very interested in the search for the missing mass! Other superparticles, according to the equations, might have mass only a little greater than that of the W and Z, and there were claims in 1985 that the CERN laboratory had found traces of squarks. But no confirmation of this was ever produced. The strongest reason for continuing to study supersymmetry is still its theoretical attraction – its mathematical simplicity and beauty. And the strongest version of supersymmetry in its original form is a variation incorporating gravity, which is called, logically enough, supergravity.

Supergravity has one powerful attraction for the theorists. As far as anyone has yet been able to ascertain, it does not fall foul of the infinities that plague lesser theories. There is still the snag that there are several versions of supergravity (eight of them, but no more), but there is one variation on the theme, called $N = 8$, which has an appealing mathematical simplicity – *if* it is set up in eleven dimensions, rather than the usual four (three of space and one of time) of the everyday world! Things were beginning to get awfully confused, with various partial theories seeming to shed some light on the problem, but each of them clearly missing some vital point, when the next revolutionary development came on the scene. That revolution is far from over, but it is causing more excitement in the world of particle theorists than anything since quarks. It is based on the idea of string.

Another way of looking at the search for a unified theory of physics is in terms of the two great theories of twentieth-century science. The first, the general theory of relativity, relates gravity to the structure of space and time. The second, quantum mechanics, describes the behaviour of the atomic and subatomic world, and there are quantum theories which describe each of the other three forces of nature, apart from gravity. A fully united description of the Universe and all it contains (a "theory of everything", or TOE) would have to take gravity and spacetime into the quantum fold. That implies that spacetime itself must be, on an appropriate very short-range level, quantised into discrete lumps, not smoothly continuous. String theory, in its extended form called superstring theory, *naturally* produces a description of gravity, out of a package initially set up in quantum terms.

The central idea of all string theories is that the conventional picture of fundamental particles (leptons and quarks) as points with no extension in any direction is replaced by the idea of particles as objects which have extension in one dimension, like a line drawn on a piece of paper, or the thinnest of strings. The extension is very small – about 10^{-35} of a metre. It would take 10^{20} such strings, laid end to end, to stretch across the diameter of a proton. The idea was introduced twenty years ago to provide a new, and possibly better, description of some specific interactions between particles. At that time, it was not thought to be a universal solution to the puzzles of physics, and although some mathematical physicists dabbled with it through the 1970s, this was as much out of interest in the mathematics itself as in any hope that strings provided a realistic description of the particle world. String theory only took off in the middle of the 1980s, when it turned out that by combining the idea of supersymmetry with strings to make a new and improved version, superstring theory, a seemingly powerful and complete description of everything could be produced.

Gravity *must* be included in superstring theory, and arises naturally in a way which can be portrayed in simple physical terms. A piece of string has two ends, and it can vibrate or rotate. The simplest motion is a rotation about its centre, with the free ends of the string zipping round at the speed of light. In string theory, such open strings corres- pond to entities in the particle world, and the properties which physicists call "charges" are tied to the end points of the strings – this may be electric charge, if we are dealing with electromagnetism, or the "colour" charge of quarks, or something else. When two such strings collide, they may join together at their end points to make a third string, and this can then split apart to make two new strings. The equations which describe this behaviour are just the family of equations needed to describe how funda- mental particles interact and scatter off of one another. But a string can also take on a different form.

Suppose the two ends of the string join together to make a closed loop. A closed string is fundamentally different from an open string, but any theory which contains open strings must also contain closed strings. When string theory is set up to describe the three forces of nature that have already been given a quantum description, the closed loops that emerge automatically from the theory turn out to have just the properties required to provide a description of gravity – they are gravitons, the carriers of the gravitational force.

All this was exciting enough to encourage more theorists to begin working on strings and superstrings. The cottage industry developed into a full-scale attack on the problems of this branch of mathematical physics when more bonuses – and some peculiarities – emerged from the calculations. The major peculiarity of the models is that they only work in ten dimensions (nine of space plus one of time). Since we live in a four-dimensional spacetime, this might seem like a serious drawback. But mathematicians have a trick, which they call compactification, for getting rid of surplus dimensions.

Think of a piece of real string, in the everyday world.

Unlike the strings of superstring theory, it has a definite thickness – it is a solid three-dimensional object. But if you look at it from far away, it looks like a one-dimensional line. The same thing would happen with a hosepipe, which has the important difference from our everyday string of being hollow. Viewed from afar, it looks like a one-dimensional line. Up close, we can see that it is really a multi-dimensional surface, rolled up on itself. Mathematicians can get rid of the six extra dimensions of superstring theory using a similar trick, requiring them to be compactified, rolled up into the multi-dimensional equivalents of tiny spheres, cylinders or circles, that would reveal their inner structure only if examined at distances much smaller than anything we could ever hope to probe.

This idea itself leads to new and fertile fields for the theorists to plough. How and why did the six dimensions get rolled up? Is it possible that compactification provided the mechanism which began the powerful expansion of the Universe in the big bang? Nobody knows. But all this work is taken very seriously because of the great benefit that emerges from superstring theory in ten dimensions. There are many possible geometrical descriptions of such theories, but just two of them do not contain any of the infinities which have plagued attempts at developing a satisfactory GUT. Since one of these geometries provides no distinction between left and right,* there is, in fact, only one version of superstring theory that contains no infinities and seems to provide a description of how the world works. This unique model emerges from very simple basic ideas, and is essentially determined only by the requirement that it should be self-consistent (which seems a modest enough requirement); it is therefore highly regarded by many mathematical physicists as the front runner in the race to develop a theory of everything.

Mathematicians are now very much involved in the new

---

* Only the weak force of nature actually does distinguish left from right in particle interactions, and it does so only under rare circumstances. This "handedness", or chirality, is a very weak feature of nature at this fundamental level. But it does exist, and must be included in any good theory.

physics. Movement of points through space and time can be described in terms of lines traced out by the particles – trajectories, or world-lines. Moving strings, however, trace out sheets and surfaces in spacetime, which require a quite different mathematical treatment. Multidimensional topologies, which some mathematicians have studied for their intrinsic abstract interest, suddenly turn out to have practical relevance.

The particle physicists get in on the action too. It is the way the extra dimensions of space get curled up that allows for the existence of a whole family of particles which interact with the known world only through gravity – the "shadow matter". And it is at least intriguing that although the existence of shadow matter is not inevitable in superstring theory, the possibility that allows for the existence of shadow matter also requires the existence of the axion, a particle which physicists already needed in order to tidy up some inconsistencies in QCD, and which astronomers would be more than happy to have as part of the missing mass. There is, it seems, something for everybody in superstring theory – even those with more philosophical inclinations are intrigued by the way in which many features of this approach are improved using the "many worlds" or "sum over histories" approach to quantum mechanics, in which the "real" world is seen as an averaging over all possible worlds.*

Superstring theory can, it seems, be all things to all physicists. The prospects are that we are in for a period of rapid development and intellectually stimulating new ideas. But there are voices crying out against the chorus of approval for these new ideas. A few theorists are worried that the young generation taking up string theory so enthusiastically may all be rushing up a blind alley, and that hardly anybody is working in other areas which may seem less exciting today but may turn out to be important tomorrow. Where would we be now, after all, if everyone had ignored string theory when it was unfashionable? But no doubt a previous generation said the same thing about

---

* I described Many Worlds cosmology in my book *In Search of the Big Bang*.

relativity theory, or quantum mechanics. However good a description of reality superstring theory turns out to be, it is already changing the way physicists think, and that must be to the good, for one very important reason.

This new development epitomises what has happened to fundamental science in recent years. There are two profound differences between the development of superstring theory and the development of other key ideas. The most important is that now and in the foreseeable future there is little likelihood of new experiments which can either test the predictions or throw up puzzles for the theorists to chew over. As physicists probe to smaller length scales and higher energies, experiments become more costly and time-consuming. The search for the W and Z particles, for example, involved hundreds of physicists and many years' work, in a laboratory so expensive that it has to be funded by several European nations in collaboration. One result as significant as that in a generation may be the most physicists can hope for in the future. So it is natural that physics, less tightly tied to experiment than it has ever been, will develop with the aid of pure mathematics – including ideas of topology and multidimensional spacetime – with philosophical overtones. Whether you think this is good or bad depends on your personal prejudices; it will certainly be different from what Newton, or even Einstein, was used to.

The second feature of superstring theory that makes it difficult to make progress is the lack of any deep insight into what it all means. Einstein used to tell how he was sitting at his desk in the patent office in Berne one day when he was suddenly struck by the thought that a falling man would not feel the force of gravity. This penetrating insight, later dignified as the principle of equivalence, led him to the basis of what became the general theory of relativity. The logic was clear before the details were worked out. In superstring theory, though, even one of the pioneers, Michael Green, has said: "Details have come first; we are still groping for a unifying insight into the logic of the theory. For example, the occurrence of the massless graviton and gauge particles that emerge from superstring theories appears accidental and somewhat mysterious; one would like them to emerge naturally in a theory after the unifying principles are well established."*

Whatever else it has done, though, superstring theory has fired a generation of physicists with enthusiasm and has brought a different kind of mathematical approach to the problem of understanding the nature of everything. The best practical hope of linking these ideas in to observations of matter lies in the search for the missing mass and our understanding of the Universe at large.

* *Scientific American*, September 1986, volume 255, number 3, article beginning on page 44. Green is Professor of Physics at Queen Mary College in London.

BIBLIOGRAPHY

The books mentioned here provide useful background information on the topics I discuss in the earlier chapters of the present book. Many of the ideas discussed in later chapters, however, are so new that they have not yet found their way into the pages of the textbooks, let alone into more accessible "popular" accounts. I have therefore, wherever possible, mentioned in the text the names of the people involved in this new research, and the places where they work. The best way to keep up to date with the new developments is through the pages of magazines such as *New Scientist*, *Science News* and *Scientific American*, where those names crop up frequently in reports of work relevant to the search for the missing mass and the ultimate fate of the Universe.

Atkins, Peter. *The Second Law*, New York: Scientific American/ W. H. Freeman, 1984.
A well-illustrated and non-mathematical introduction to the supreme law of nature, the second law of thermodynamics.

Barrow, John and Frank Tipler. *The Anthropic Cosmological Principle*, Oxford University Press, 1986.
An enormous book (over 700 pages) in which the authors

digress on just about every subject relevant to the theme of mankind and the Universe, including a good, detailed section on its ultimate fate. Quite technical in parts, but an absorbing read in others. Good to dip into in the library!

Davies, Paul. *Space and Time in the Modern Universe*, Cambridge University Press, 1977.
A delightful little book, which covers relativity and distorted spacetime, black holes, the traditional interpretation of thermodynamics and the relationship between life and the Universe, all in the space of 222 pages. Very well written; only slightly out of date.

Eddington, Arthur. *The Nature of the Physical World*, Cambridge University Press, 1928.
Although written sixty years ago, Eddington's book is still worth reading for his discussion of the basic philosophy of the scientific approach and his still clear explanation of concepts such as the thermodynamic arrow of time and the gravitational curvature of space. The quote on the title page of the present book is from page 74 of this edition.

Flood, Raymond and Michael Lockwood (editors). *The Nature of Time*, Oxford: Blackwell, 1986.
A collection of essays based on a series of public lectures in Oxford in 1985. Slightly patchy, but with very good contributions from Paul Davies, Peter Atkins and Roger Penrose on themes related to the fate of the Universe.

Gribbin, John. *In Search of Schrödinger's Cat*, New York/London: Bantam/Corgi, 1984.
The story of the development of quantum physics – the astonishingly successful, but non-commonsensical, theory of the world of the very small – during the twentieth century. Particle-wave duality, quantum uncertainty and the rest explained in (I hope!) readable fashion.

Gribbin, John. *In Search of the Big Bang*, New York/London: Bantam/Corgi, 1986.
In my biassed opinion, the best up-to-date account of modern ideas about the origin of the Universe and the interface between particle physics and cosmology.

Harrison, Edward. *Cosmology*, Cambridge University Press, 1981.
    Really a textbook, but still accessible to the general reader. The best place to go for a broad overview of modern ideas about the Universe.

Kaufmann, William. *Universe*, New York: Freeman, 1985.
    A nicely written and well-illustrated guide to astonomy, but one which, for my taste, puts too much emphasis on stars and planets and not enough on the Universe at large. Designed for use in teaching a general science course, and worth dipping into if you can find it in a library. Shu's book (see below) is better.

Narlikar, Jayant. *Introduction to Cosmology*, Boston: Jones and Bartlett, 1983.
    The best no-punches-pulled textbook for anyone with a serious interest in cosmology and a thorough grounding in mathematics.

Ne'eman, Yuval and Yoram Kirsh. *The Particle Hunters*, Cambridge University Press, 1986.
    The story of particle physics from the discovery of the electron and the proof that atoms are divisible (almost exactly a hundred years ago) to the evidence for the W and Z particles, in the 1980s, that suggests that today's physicists are on the trail of a unified theory linking all the known particles. Just the place to find out about quarks, baryons, leptons and neutrinos.

Prigogine, Ilya and Isabelle Stengers, *Order out of Chaos*, New York: Bantam, 1984.
    Nobel laureate Prigogine made his name in the 1970s with new ideas about the meaning of the laws of thermodynamics and the nature of the Universe. This semi-popular account of his work provides the most complete and up-to-date explanation of thermodynamics for the non-specialist, and includes discussion of the arrow of time. A serious and deep book, which rewards careful reading. For those in a hurry, though, Prigogine's ideas are summarised in snappy fashion in Alastair Rae's book, mentioned below.

Rae, Alastair. *Quantum Physics: Illusion or reality?*, Cambridge University Press, 1986.
A readable little book on the puzzles of the physics of the world of the very small – atoms and particles. Mainly, as its title suggests, quantum physics; but including an excellent chapter on thermodynamics and the arrow of time, with the best concise summary I have seen of the latest ideas from Ilya Prigogine.

Rowan-Robinson, Michael. *The Cosmological Distance Ladder*, New York: Freeman, 1985.
The best and most accessible explanation of the way astronomers determine distances to remote objects, and thereby get clues to the age of the Universe and its ultimate fate. For a closed universe with omega equal to one, we need a rather lower value of the important Hubble parameter than Rowan-Robinson favours as the most likely value, but at the time he wrote this book he was not aware of the implications of the work on "WIMPs" discussed in my Chapter Seven.

Shu, Frank. *The Physical Universe*, Mill Valley, California: University Science Books, 1982.
A great, big, delight of a book, aimed at liberal arts majors, which gives a solid overview of the Universe and our place in it for those who lack the mathematical background to appreciate, say, Narlikar's textbook.

Silk, Joseph. *The Big Bang*, New York: Freeman, 1980.
Written at a slightly more technical level than the present book, but still a very readable exposition of the basics of cosmology and the reasons why astronomers think the Universe began in a big bang.

Teilhard de Chardin, Pierre. *The Phenomenon of Man*, London: Collins, 1959.
A sometimes sticky exposition of Teilhard's philosophical ideas about Christianity and the evolution of consciousness. Even Sir Julian Huxley, in the Introduction, says "his thoughts are not fully clear to me", so don't feel too embarrassed if they aren't clear to you either! Notable, though, for the introduction of the term Omega Point for the ultimate state of the Universe.

INDEX

ABOUT THE AUTHOR

Dr. John Gribbin, science writer and cosmologist, is the author of TIME WARPS, WHITE HOLES, THE REDUNDANT MALE, GENESIS, THE MONKEY PUZZLE, IN SEARCH OF THE BIG BANG, IN SEARCH OF THE DOUBLE HELIX, and the best-selling IN SEARCH OF SCHRÖDINGER'S CAT. He holds a doctorate in astrophysics from Cambridge University and lives in East Sussex, England.